next century demands. Read this book, it affirms our humanity, and the civilizational genius, and the imperative for synergic relationships between peoples and flowing water—common to all languages, climates, and cultures. And it's a joy!"

—Severine Von Tscharner Fleming, cofounder Agrarian Trust, Greenhorns, Farm Hack, Smithereen Farm, publisher of *The New Farmers Almanac* vol 1–7

"A vital book for people and policymakers alike. Jain and Franses remind us that our intimate connection to the water cycle is, and always has been, hyper-local. The days of big, concrete, centralized systems are over—we need to think locally, capturing and storing rainfall and stormwater where it falls, with nature-based solutions. Through years of on-the-ground projects across several continents, Jain and Franses give us the examples we need—from rural Rajasthan to cities in Slovakia—to secure our water future and gain cooler, greener local climates at the same time. The example of two Indian villages 30 kilometers apart—one 5 to 6 degrees cooler than the other just through good water capture and conservation practices—will blow your mind! This is the phrasebook we all need for climate adaptation and water security—the good news is it is a common language we all once knew, and can easily relearn."

—Tim Smedley, author of *The Last Drop: Solving the World's Water Crisis*

"*The Language of Water* is an important, inspiring and visionary book that offers great hope and far-reaching implications for the future. At its heart is the much-needed transformation of our relationship with nature and particularly the water cycles of the planet we all share. The authors share astonishing stories of transformation, and commonsense solutions that have grown out of the deep wisdom of learning from time-honored traditional solutions, of listening to the land, and valuing the people who have a long-standing relationship with their land.

The book demonstrates how small-scale community actions have far reaching consequences, both within the communities themselves and as part of the bigger picture as we shift our awareness towards an understanding of the deep-rooted interconnectivity of nature's cycles. It is an uplifting read for all those who are intimately connected to the land."

—**Glennie Kindred, author of** *Walking with Trees*

"*The Language of Water* opens our eyes to the longstanding interplay between humanity and the Earth's water cycle. Authors Minni Jain and Philip Franses examine low-tech water-management techniques, some in continuous use for millennia, others rediscovered only recently following disastrous industrial interludes. In their on-the-scene reporting from the UK to India to Zimbabwe to China to Colombia and beyond, we see communities preventing floods, reading the desert landscape to locate groundwater, reversing colonial-era land and water abuse, maintaining a 2,500-year-old irrigation system, or rescuing a vital water source ruined decades ago by deforestation. They even drop in on a sophisticated water-management project carried out entirely by beavers. In all of this, Jain and Franses show us how our communities can avoid treating nature as 'a passive receptacle of already structured solutions' and instead let nature be our teacher and partner as we learn once again to speak the language of water."

—**Stan Cox, author of** *The Green New Deal and Beyond: Ending the Climate Emergency While We Still Can*

"*The Language of Water* is a powerful book. It is an informative as well as an inspiring book. The authors have woven together a groundbreaking narrative of examples and stories which show that committed activists can create miracles! In the context of climate change we need to take care of water, the most precious source of life! Read this book and discover how individuals and communities around the world are addressing the challenges of water conservation. This is a book of hope and a book of action; a book which might make you a water activist!"

—**Satish Kumar, president Emeritus, Schumacher College**

"The language of water demands a school to resource its speakers, to communicate its stories, to spread its simple message. The authors of *The Language of Water*, founders of The Flow Partnership, delve into the precar-

Praise for *The Language of Water*

"'Water' and 'survival' are pretty much the same thing, so it's no wonder that local communities, facing record drought and heat, are taking matters into their own hands. These are stirring stories of the recovery of time-honored techniques that will be desperately important as the climate crisis keeps building."

—Bill McKibben, author of *The End of Nature*

"Finally! A book—a brilliant book—that explains the crucial role of water in maintaining a healthy climate and preserving life on our planet. Using ancient knowledge applied to our current crisis, the authors show us how to understand and protect the hydrologic cycle and how to 'hold' water in order to green deserts, bring back life to drought-stricken lands and manage the all-too-frequent floods plaguing our cities.

The language of water is something every one of us understands at the deepest level of our being, but that has been lost in our modern world. This wonderful book shows us how to recover the language of water if we are to save Mother Earth and ourselves."

—Maude Barlow, activist and author

"This extraordinary work demonstrates many times over that there is no power for change greater than a community working for what it cares about. Working with life's dynamics and engaging the human spirit reliably leads to astonishing results. Beautifully written and illustrated, I want everyone engaged in community work to read this book."

—Margaret J. Wheatley, author of *Leadership and the New Science* and *Restoring Sanity.*

"Water is the physical manifestation of the Oneness of Humanity on our planet. The water that flows in the River Thames, outside the Houses of Parliament, is One with the rain that falls in arid Rajasthan, India, and also One with the torrents that flood the cities and plains of Europe. Every stream, river, and catchment across the world, whether in a state of flood or drought or stability, is an expression of the Oneness of water and connects us Humans to this Oneness.

With community stories from across the world, *The Language of Water: Ancient Techniques and Community Stories for a Water Secure Future* is a book unlike any other. It gives us a real and achievable route to climate stability and abundance by holding water in the planet's landscapes. In this Indigenous knowledge and wisdom are as important as modern science and technology. By giving local communities the wherewithal to hold the water in their regions, we are well on the way to taking real, large-scale action to rebalance the climate and address at source many of our related, wicked problems of hunger, poverty, and migration."

—Andrew Stone, Baron Stone of Blackheath

"*The Language of Water* is a love story about humans and water, and the ways in which we are learning to keep this relationship healthy for the sake of future generations, reminding us that it takes people in our local communities and those at the macro level in cooperation to help care for Mother Earth's waters. With cases from around the world as evidence, Minni Jain and Philip Franses show us that for the sake of Mother Earth's future, community-led initiatives and traditional wisdom partnered with earth-based scientific practices and research are the path of hope. Sustainable relationship with the waters around us is essential if we are going to act in order to give future generations a chance—it's all connected through kinship, community, action and care. Read this book as a reminder of the love story that is always unfolding, and the important part we have to play in it."

—Kaitlin B. Curtice, Potawatomi poet-storyteller and
award-winning author of *Native* and *Living Resistance*

"The directive is clear: follow the logic of the landscape. This kind and clarion book leads us on a learning journey through the planet's many vernacular approaches to restoration hydrology. To participate in our survival as a species, the work we must pursue is in healing our relationship with land and water. Turns out, the rearrangement of elements along a gradient drives as a living metaphor and social practice so much of what our iousness of our environmental health and water security through stories of communities and individuals across the globe and in nature itself. In doing so, their book becomes a perfect textbook for the universal school they propose. *The Language of Water* is a worthy read, complete with case studies and data that underscore the importance of sensible water management."

—**Nancy Castaldo, author of *When the World Runs Dry***

"The authors chronicle areas in India, the UK and Africa where the scarcity of water is critical. Water is life, without it all beings cannot exist. Climate change has proven how precious water is. It can no longer be looked at as an infinite resource that will last forever, but a resource that must be preserved.

Yet in drought-stricken countries, millions of gallons of rainwater are wasted. Throughout the book the need to capture and preserve water is presented as a universal language which everyone can understand. You don't need to be a rocket scientist or have a degree in hydrology to know water can play a major role in mitigating climate change. The book demonstrates the importance of ancestral knowledge for various communities who have treated water as a life force, understanding its ebb and flow. As the authors state, the language of water is not a human construct but a language in which water preservation is the responsibility of all."

—**Karen Washington, farmer and activist**

"*The Language of Water* is a very important and inspiring book. It is filled with examples around the world where people take responsibility for healing their landscapes and improving their lives. They use a wide variety of solutions, guided by the imperative to hold the rains from washing away, enabling the water to rehydrate the earth, reviving soils, even springs and rivers. In some places, traditional techniques which had fallen in disuse were revived; in others, new solutions are found. Everywhere these communities work from their specific ecological contexts, reversing desertification and the poverty it engenders. Sometimes, it is the women who take the lead, since they are the ones who suffer most from lack of water resources. In the process, they redress gender inequalities while enriching their villages and regions.

The authors argue passionately and persuasively that meeting climate change challenges and positively charting the future starts from and depends on grassroots action. *The Language of Water* is an empowering book—a clarion call for taking the actions to heal our lives and our world, starting with respecting and honoring water and its power to nurture life."

—Mark Nelson, PhD, chairman, Institute of Ecotechnics, and author of *The Wastewater Gardener: Preserving the Planet One Flush at a Time* and *Life Under Glass: Crucial Lessons in Planetary Stewardship from Two Years in Biosphere 2*

"*The Language of Water* is an amazing book filled with examples of traditional water management from many parts of the world that demonstrate solutions to our drought and flood-prone landscapes. The inspiring stories of communities making water containment structures and taking other measures into their own hands show us that despite slow action from 'the powers that be,' we can all be involved in the restoration of a healthy water cycle and all that flows from it."

—Martin Crawford, author of *Shrubs for Gardens, Agroforestry and Permaculture* and founder and director of The Agroforestry Research Trust

"Poetically called *The Language of Water*, this book shares a sweeping overview uniting various perspectives of water—poverty alleviation and large water projects, commodification and scientific and financial perspectives, and more. What is most valuable, however, is the painstaking work done by Minni Jain and Philip Franses in connecting with and sharing stories of the individuals and communities around the world who have worked to talk a sane and beautiful language of water—providing food and a sustainable way of life. Huge projects—be it skyscrapers, dams, and canals or monocultural, large-scale, water-guzzling agriculture—have become the norm for colonial powers and modern governments today. But what comes out clearly in this book is that it will be the diverse, local communities combining modern and traditional wisdom that can solve the complex water problems of the world; the dams of logs and branches and natural flood-management techniques should not be sneered at by science and technology but valued appropriately where they are the more effective alternative.

The authors make the book interesting reading for the layperson as well. From looking at beavers as hydrologists to the concept of local water banks, from the yin and the yang of gradual shifts from male dominance to equality for women and the Water Schools of The Flow Partnerships—in multiple ways the book brings valuable insights into the socioeconomic realities of life today and the solutions we need.

As they say in their closing chapter, 'Never underestimate the power of persistent community engagement, even in the face of hopelessness. *If all we have is our imagination, voice, and action—let us not be afraid to use them!*'"

—Seetha Ananthasivan, founder, Bhoomi College, and Prakriya Green Wisdom School, Bangalore, India

"The Greek philosopher Thales, the ancient Hindu scriptures, and the songs of the Amazon all echo one spiritual truth—water is the essence of life. We all are born from it, and water transforms itself to become human bodies, trees, food, animals, and the world we see around us. *The Language of Water* captures this journey and beautifully narrates the transformation of our divine water across human cultures, delving into how diverse cultures have related to water and retained their sovereignty. This book is a much-needed global water encyclopedia. As the water crisis is already upon us, this handbook offers insights and solutions as well as a deep study of indigenous and industrial approaches. The triumph of the indigenous systems against newer challenges presented by both our changing climate and predatory exploitation are also well documented here. Readers can familiarize themselves with real, positive indigenous examples based on decentralized, community-driven water conservation systems. I congratulate the authors of the book and hope all water activists, students, and concerned citizens get a chance to understand our planetary water systems through this work."

—Indra Shekhar Singh, writer and host of *Krishi Ki Baat— Farm Talks* for *The Wire* and former director of policy and outreach for National Seed Association of India (NSAI)

THE LANGUAGE OF WATER

*Ancient Techniques and Community Stories
for a Water Secure Future*

Minni Jain and Philip Franses

Foreword by Pascale d'Erm

SYNERGETIC PRESS
SANTA FE • LONDON

Published by Synergetic Press
1 Blue Bird Court, Santa Fe, New Mexico 87508
& 24 Old Gloucester St., London WCIN 3AL, England

The Library of Congress Cataloging-in-Publication Data is available from the Library of
Congress.

ISBN 978-1-957869-19-3 (paperback)
ISBN 978-1-957869-20-9 (ebook)

Cover and interior design: Jonathan Hahn
Cover illustrations: HelloRF Zcool/Shutterstock (Hand); suns07butterflyl/Shutterstock
(Blue circle)
Production Editor: Allison Felus

Printed in the United States of America

To all those communities and people, known and unknown, who hold water in their landscapes, looking after their regions without seeking any recognition and who do it simply for love of life and love of the planet. They teach us the power of working with nature and keeping our planet healthy no matter who we are, where we are, and how many or how few resources we have. Speaking the language of water is not difficult when we take self-responsibility for learning and practicing it. These communities show us how, and this book is their voice.

All royalties from this book will go back to local communities to build their ponds and hold water in their landscapes.

CONTENTS

Foreword by Pascale d'Erm xv
Acknowledgments xviii
Introduction xxi

1 Why Speak the
Language of Water? 1

2 Rajasthan, India:
A Three-Day PhD in
the Language of Water 15

3 China: Dujiangyan 32

4 Language Basics 45

5 United Kingdom:
A 5 Percent Future 61

6 Africa Unfiltered 84

7 Zimbabwe: The Spirit of
the Land Reawakens 104

8 Chimanimani, Zimbabwe:
Community Guardians of
Water 116

9 Burkina Faso: Transforming
the Desert 131

10 Slovakia: Building
a Rain Garden 146

11 Qatar: Regreening
the Middle East 158

12 Australia: Dreamtime
Down Under 172

13 Bundelkhand, India:
Making Friends with
Water 187

14 United States/United
Kingdom/Europe:
Beavers, Nature's Water
Engineers 206

15 Colombia, South America:
Steps to Community
Action 217

16 The World Water Bank 234

Conclusion 249
Notes 263
Bibliography 279
Subject Index 286

FOREWORD

Ripples, cycles, droplets, river bends, whirlpools, meanders, flows, eddies, currents (visible or invisible); expansion, evaporation; the cool rush of a mountain stream; the rising sap in trees; a circulation in our bodies . . . water weaves in and out, rises through capillary action, and evaporates before falling again as rain. From this constant movement emerge ceaseless transformations. We are linked to water in the present and also share the history with it, which takes us back to the origins of life.

A new order is preceded by chaos, says the Tao Te Ching. In these times of instability and change, the nimble, water-inspired solutions presented in this sensitive and scholarly book are a major source of inspiration, enabling us to imagine alternative ways to reforge our links with nature.

If "water splashed on a face revives the power of vision," as the poet Gaston Bachelard writes in *Water and Dreams*, then this book offers us a fountain of fresh water to do so!

As water stewards, Minni and Philip give us access to ancient wisdoms revived by communities worldwide, to foster water's return to habitable land. Like drops of water, we run down a Zimbabwean slope made fertile once again with gabions and rockpools, water a rain garden in a Slovakian school, collect thousands of cubic meters of rain in a Rajasthani johad in India before filtering slowly down into the water table below, or join the mist over the waking dreams of Australia's aboriginal peoples. Alertness and questioning gain depth when they combine with enthusiasm and poetry, which is what this book—journey—shows us.

At baseline is the denunciation of the lead weight surrounding the grip that colonialism has on local culture, which like an opaque layer painted over a precious wood, has prevented life from emerging and breathing. We are stuck in destructive and suicidal patterns of behavior that harm our well-being and the health of ecosystems because we have lost sight of the symbiosis that connects all elements of the natural world. As a result, we have also forgotten how sacred water is. So, what knowledge can we call back to mind in order to survive the ecological crises we are facing? With solutions implemented by villagers (barefoot engineers!), we experience a fundamental message: community knowledge is an answer to crisis.

The key word here is reappropriation. The reappropriation by communities of the ancestral knowledge of water and of their lands, where nature is no longer disconnected from culture, where the body is no longer isolated from its spirit or its soul, where the feminine unites with the masculine and the two polarities alternate happily. Water is a quest for equilibrium, even if precarious or unstable. This is a discerning equilibrium, where every plant, animal, mineral, and human has a role to play. The need to maintain and preserve water is extreme in arid lands devastated by climate change and requires communities to be innovative, creative, and daring.

I met Minni in the thirsty landscape of Rajasthan. As a film director I was writing a documentary series for the Arte TV channel, called *Soeurs de la Terre*, about the women involved in the regeneration of their lands. After many months of discussions, we had agreed to go and meet the Karauli communities mentioned in the book, until a few lines in a document, mentioning the existence of the Jal Sahelis, decided otherwise. Wearing turquoise saris, these women, the Jal Sahelis (Friends of Water) who revive water's sacred power by healing the human and social fabric as well as fighting for equality, were astounding enablers of life and hope! Minni kindly agreed to share the vibrant story of these village women from Bundelkhand in middle India, who manage to push back the heavy burden of patriarchy and take the matter of bringing the water to their villages into their own hands. During their remarkable pani panchayats (water councils), decisions on how to administer and manage the water basins, pumps, and wells are followed by admirable dance flourishes and expressions of joy with music! Filming was an adventure whose intensity was beyond words. We were enthralled by these water warriors. They work with and on behalf of water, and nothing will ever stop them in their struggle. By revivifying the ancestral knowledge of johads, they resanctify water, honor life, and make their lands and regions abundant again—all with a selfless confidence and joy in their work.

Our civilization lacks imagination as far as water is concerned. Although it is rare and precious, water is too often seen as a liquid resource that just comes out of a tap. But for how much longer? In *A Sand County Almanac*, Aldo Leopold notes, "We can be ethical only in relation to something we can see, feel, understand, love, or otherwise have faith in." Whether it is by rediscovering a more profound link with water, renewing our imagination, or equipping ourselves with a box of tools and grounded ways of bringing water back into its primary place, this book offers a multitude of solutions to heal, soothe, and regenerate nature and reforge a deeply emotional and imaginative connection with water.

Thanks to water's intrinsic virtues, this notion of interdependence becomes a discerning experience from which a deep awareness emerges. "From space, we can see and study the Earth as an organism whose health depends on the health of all its parts. We have the power to reconcile human affairs with natural laws and to thrive in the process," said Gro Harlem Brundtland in 1987 in *Our Common Future*, the first UN report on sustainable development.

Water teaches us a way to ascend to a higher level of human consciousness. The stories of the different communities speaking the language of water guide us with a dose of pure awareness. From science to love, commitment, and awareness, the solutions implemented and recounted throughout this book remind us that water connects us to a timeless memory. Water is the harbinger of what connects continents, histories, and geographies. By following the cycle of stories presented in this book, we become one with water.

Water restores power, wisdom, intuition, fertility, creativity, health, and longevity in those who know how to forge an alliance with it.

May you read this precious book drinking fresh water, listening to the sound of rain or a river, enthralled by the translucent blue of a mountain lake, or transfixed by ocean waves. Offer it as a gift, for I wish for this book's messages to flow in abundance.

Pascale d'Erm
Quimper, Brittany, France
July 17, 2024

ACKNOWLEDGMENTS

We would like to acknowledge all the communities engaged in speaking the language of water around the world, without whom this book would have been nothing but a series of depressing statistics or never ending "evidence" around increasing floods and droughts.

Our gratitude to those dedicated water practitioners who have pioneered the restoration of water with communities around the world and kept their work alive and now increasingly relevant: Rajendra Singh and his team for communicating the imperative of communities taking action to restore their arid lands, reviving numerous rivers in India; Michal Kravčík, Danka Kravčíková, Peter Bujnak, and others at People and Water International, Slovakia, for assessing the task of rehydrating the land at a global scale and engaging communities nationally to realize this; Ziwei Fan, Long Xiang, and Master Peng for introducing us to the echoes of this language of water from 2,500 years ago as in Dujiangyan, China; Paul Quinn, Mark Wilkinson, and Marc Stutter of the James Hutton Institute for their ceaseless support for natural catchment restoration and making it fashionable to speak this community language of water in the UK; Mark Fletcher, Dervilla Mitchell, David Hetherington, and Louise Bingham from ARUP for showing us the bridge between engineering and community; Harjinder Sembhi and Rajiv Sinha for crossing the bridge between science and community; our Water School Africa partners Abraham Ndhlovu, Daniel Ndhlovu, and Emmanuel Prince of the Muonde Trust, Walter Nyika Mugove of ReScope Zambia, Peter Gubbels and Tsuamba Borgou of Groundswell International, and Zachary Makanya of RIDEP who have been speaking this language tirelessly in the villages across their continent; John Wilson, who has been ceaselessly putting agroecology principles in place with local communities in Zimbabwe and who introduced us to some of the magnificent water retention that is already happening in projects like CELUCT, Nyahode Union Learning Centre, PORET, and ZIMSOFF; Felipe and Alejandra Medina for learning the language of water anew through the children in Barichara, Colombia; Sanjay Singh, Pratiksha Tripathi, the Jal Sahelis Sri Kunwar, Neelam Jha, Phoolvati (and the 1,500-strong others) who have overturned traditional norms to equip women to lead the work of restoring the water and landscape in their villages and managing them wisely; Andrew Stone, who swung his might behind this work when it was still fledgling and finding its voice and continues to do so; Seetha Ananthasivan and the team at Bhoomi College Bangalore who have chased and made real a vision of ecological and spiritual living that is in complete harmony with nature showing us how a micro water cycle can be kept functioning and alive; Pascale d'Erm and Christoph Schwaiger, who understood the power of the work of the Jal Sahelis and the women leading the change that is needed in speaking the language of water and who told the story through film, helping it become a global message; and to Ann Topping in bringing out the beauty of thoughts and words in translations.

And there are so many who have added to, endorsed, and supported this work in many, many different ways over the years. Pupak Hagigi and Alan Featherstone in Scotland and Susy Patchett Williams and Nick Batt in Devon shared their place of beauty most generously and warmly so that we could write these stories. We are also grateful to Nicole Franses for the countless times she has allowed us into her space to work and speak with Africa, India, and the rest of the world.

To all nature and the beavers who are part of it, who impart lessons in speaking the language of water to us: thank you.

To all those funders, companies, organizations, known and anonymous, who have the vision to resource this work: thank you.

And finally at Synergetic Press, our gratitude and deepest appreciation to our editor Noelle Armstrong, who first proposed we write this book and tell the stories of the communities who are repairing the planet and then guided us sensitively and expertly throughout the writing of it. We can safely say if it hadn't been for Noelle patiently encouraging us to tell these stories, this book would not have been written! A special thank you to Deborah Snyder and Allison Felus for holding all the intricacies of what it takes to bring this book to the reader. Also to Jasmine Virdi, Doug Reil, and the rest of the dedicated team for their support. This is a team endeavor, and Synergetic Press is a publisher that still cares about what we can do for the planet and works tirelessly to give that message to those who want to listen.

To all of you and those we may have missed: thank you.

The language of water is timeless and will never become extinct as long as we keep it alive and spoken amongst the small, ground-level communities of this world.

INTRODUCTION

If we imagine restoring our planet to health, where does the real power to do that lie? Is it found in the towering skyscrapers and ever-expanding sprawl of a city, or in the small, rural communities throughout the world who remember how to hold water in the ground and sustain an abundant land with lush crops and trees? Is the current climate instability we are experiencing just a top-level policy predicament? Or is a solution already present in communities and their traditional methods of holding water? These questions led us to found the Flow Partnership in 2011 with the aim of creating a global yet decentralized network of local communities sharing their wisdom and knowledge and rebalancing the water cycles on our planet.

As we began visiting, researching, and documenting how communities were rejuvenating landscapes, the understanding that this could happen at a large enough scale for rebalancing the climate came more and more sharply into focus. Could soft engineering solutions form the basis of a new paradigm of water holding as a strategy to renew landscapes around the world?

This question seems to have been answered at Bhoomi College, a beautiful, verdant ecological studies college located just outside the megacity of Bangalore in India. Over the years Bhoomi has planted many trees and the campus is rich in flora and fauna, with a diverse variety of birds and colorful butterflies fluttering throughout the grounds. In addition to their own kitchen garden, a few small farms nearby supply almost all of Bhoomi's food. Many people come there to learn and teach their wisdom in inner and outer ecology. When we taught and participated in a course on holistic natural systems on our first visit there in

2015, the college, though located on the outskirts of the city, was still situated in a rural location. The lazy, winding village road leading to it passed patches of woods interspersed with quiet villages where even the stray dogs barely flicked a glance at us. Today, a chaotic skyline of messy, high-rise tower blocks has gradually encroached upon that space. These days, the glossy signs announcing new residential developments (with names like Walnut Creek, without a walnut tree or creek in sight!) are now within touching distance, nibbling away at Bhoomi's doorstep. In 2024, that calm unhurriedness we had encountered has been overwritten by a busy, congested through road connecting it all to the city.

Not a single thought is being paid to where the water for those high-rise tower block residences at Bhoomi's edge is going to come from. Currently, it is being ferried in tankers that siphon the water from lakes outside the city. The city's own groundwater table has been depleted and is unable to fulfil its needs.

In the 1800s, Bangalore had 1,452 lakes and 80 percent green cover. Up to a few decades ago, Bangalore was like a cool, air-conditioned city in the heat of India. Then the development boom and the establishment of Bangalore as an IT center put a premium on housing. Nature had to make way: forests were cut and lakes began to be filled in as the perimeter of Bangalore reached ever outward. Individual greed for urbanization and its profits has reached such levels that it has triggered a collective crisis for society. Without vegetation cover and bodies of water, scant rainwater is held in the land and the groundwater is unable to recharge sufficiently. When Bhoomi College was started in the early 2000s, the bore holes to access groundwater were drilled down to only 250 feet. Now, they have to be drilled down to more than 1,000 feet and still, sometimes, are unable to reach the groundwater. The high-rises, some of which contain more than 500 flats, are dependent solely on water tankers. Water rationing has becoming the norm. Bangalore is now left with only around 193 bodies of water and 4 percent green cover.[1]

With this ever-expanding, thirsty city dependent on water tankers, which in turn are dependent on dwindling water sources outside the city, Bhoomi College feels like the last bastion demonstrating how nature can transform its environment for the better and reward us with its abundance of water and natural gifts for a healthy life.

This relentless expansion symbolizes universal issues that are gripping cities all over the world. In 2024, as we write this book, the world has moved to a very different place from where it was in when we began the Flow Partnership as an NGO in 2011. With every extreme climate event, the capacity of our engineering systems to regulate nature for our benefit approaches the point of breakdown. The inability of our institutions to cope with (or at times even acknowledge) the obvious instability of climate events happening around us is experienced as a deep cultural insecurity. Governments and institutions still cling to the idea that we can explain, control, and engineer our way out of the now, worryingly, all-pervasive and ever-present climate instability.

Climate instability further revealed itself as an undeniable catastrophe with the terrible wildfires in Australia in 2020. Suddenly, a conjunction of drought, excessive temperatures, and dried-out vegetation painted the picture of a world ruled by an unpredictable climate. We can see in ever more obvious ways how failing to store the water and cutting vast numbers of trees, releasing CO^2 into the atmosphere to unsustainable levels, has destabilized the climate. Our toolkit of engineering solutions for mastering planetary dynamics suddenly seemed totally inadequate to deal with nature that could ferociously surpass all limits.

In the cities, increasing mental stress is apparent in people living in a totally man-made environment. Meanwhile, in the countryside, where the dryness of the ground after prolonged drought or the rise of the river swollen with extreme rains are more tangible, there is a different anxiety—if people can't grow food anymore, they will have to leave their homes and migrate.

In this book, you will find the language of water expressed within the stories of communities who are addressing the degradation of their landscapes and demonstrating how to take individual and collective action, wherever we are living. The agency for change no longer lies just with the levers of power and policy. Through communities who have taken action, we can see our own agency in looking after the gifts of nature that keep us alive and healthy. In learning the language of water, we can speak coherently about climate instability and, dare we say it, begin to address it. No longer do we need to say, "Why doesn't the government do something about it?" The language of water starts with something that is already apparent: the water cycle. We do not have to reinvent it or debate its existence; we just have to relearn what it is and do what we can to play our part in keeping it functioning. Whether we live in a city, a village, or somewhere in between, we can revive our local water cycles at our own level.

By understanding how human intervention can strengthen the water cycle, community by community, a pattern becomes evident: local water deposits translate into global water availability. Actions such as using less water, holding rainfall in the ground, planting trees, and making the soil porous can go a long way toward repairing broken water cycles and allowing some stability in our climate again. We can begin to address the looming issue of climate change and global warming that so many feel powerless against. We are not powerless. Keep turning these pages to find out why.

An illustration of how reforestation can mitigate climate change can be found in the difference between the East Coast and the West Coast of the United States. When one of our colleagues contrasted growing up on the East Coast, where rain was abundant, with living and working in California, where the climate is increasingly dry, we consulted a 2024 research paper to understand why this might be the case. Barnes and associates demonstrate[3] how reforestation in the eastern United States has cooled the surface temperature over the last century. The water-deprived regions in the West see headline upon headline about their rising temperatures, now up to one degree Celsius hotter, while the eastern part of the United States, with its young forests, has actually cooled by up to 0.5 degree Celsius. (Refer to the Eastern and Western US reforestation and temperature correlation maps in the photo insert.) This is just one example of what can happen when individual communities in a local region repair the water cycle through reforestation.

Instead of responding to climate change by controlling water to maximize productivity on the land, we could enable our communities to focus on holding water and planting trees as a strategy for global cooling.

This book is about how communities speak the language of water fluently, and how you can too. Unlike other languages, the language of water is not specific to a country. Everybody speaks it, and it will take little for all of us to speak it fluently again. If we continue to commodify water and drive a synthetic economy by calculating how much each cup of water should cost, trading water so that the price rises as the resource diminishes, making a few people wealthy, what will become of us then? In this book, we are returning the language of water to an individual understanding that we can learn through our own practice. Once we speak this language, we can address nature in its own dialect. We are speaking to our shared anguish about the state of the world and nurturing that glimmer of hope that we each have the ability to effect change. As readers, as community farmers, as activists, as scientists, as policy makers, we speak the language of water together.

The stories in this book are in the voices of individuals and communities, written in their language, with their understanding of how they are taking action to redress the imbalance of a rapidly heating planet. In hearing their voices, we understand once again how local solutions can address the bigger challenge of the global climate imbalance; we become heartened to take those small steps ourselves.

Respect for nature, holding water, caring for soil—though these actions may seem small when dwarfed by the scale of the skyscrapers, they are showing us something fundamental about the underlying rules of nature on which everything depends. If we do not look after nature, we will face the ultimate price of a landscape that is worth nothing without water to serve it. Bhoomi College and other small communities like it are a test case for the future of the whole planet.

What do we learn from nature that can help the city leap into a sound future? At the center of our technology, our innovation, our scientific discovery is a transformation of our relationship to nature: how do we hold water to enable the earth to flourish? Our landscapes reflect the inner choice we make between destruction and creation.

1

Why Speak the Language of Water?

Water is the alphabet of life given to us by nature. If water is the alphabet, then the landscape is its dictionary. We speak the language of water to maintain a water balance in the landscape, otherwise we will have floods, droughts, or varying degrees of both, making the planet increasingly uninhabitable. With climate change becoming the defining and most worrying issue of our time, what are the nuances of this language of water that can help bring the climate on earth back into balance?

If we want to regenerate, restore, and rejuvenate the planet, we must hold the water that falls from the sky. If enough people do that, we will be taking care of our planet, and possibly even beginning to reverse climate change at its very basic level.

Water's role in sustaining life on earth is primary and unique and there is no replacement for it. The fundamentals of the language of water can be seen in the optimum functioning of the water cycle, which is crucial in keeping our planet's healthy climate sustained across millennia.

Water is the common shared language for life. Everyone on this planet speaks it, in our own bodies (60–70 percent), in the planet (71 percent), and in the systems that sustain us. When water starts vanishing, we see multiple breakdowns in our life systems as is evident in the climate change scenario facing us today. And the water cycle needs us to collaborate with it to keep life's systems functioning.

A comparison could be made with money in a bank. If we speak the language of moneymaking, our deposits and wealth in the bank increase. If we speak the language of water, our deposits and wealth in the groundwater, the soil, and the rivers increase and there is plenty to keep all it nourishes alive. In either case, if the deposits vanish, then life becomes difficult. There is, however, a crucial difference between these two banking systems. When it comes to money, we can always generate another system, but if the water banks on the planet collapse, life would cease.

As the quote attributed to Jacques Cousteau says, "We forget that the water cycle and the life cycle are one."

Our current situation can be visualized with the NASA temperature maps found in the color insert. The first one is from 1976, a time that seemed to be full of promise and potential. Fast-forward to 2024 and look at the colors in the world now. How did we manage to go from a salubrious state in 1976 to this angry heat in 2024? *What happened?*

Something seems to be alarmingly wrong with the water balance on the planet. Will it be able to support life if it continues getting hotter and drier at the pace of the last 50 years? (See Figure 3 in the photo insert.)

Figure 4 (in the photo insert) shows another alarming picture: the freshwater stress map of the world, depicting the amount of groundwater withdrawal against the amount of water available to the various countries in the world. Saudi Arabia uses 10 times the amount of water that's available in its own lands. Because the Middle East can no longer survive on the groundwater it contains in its lands, it is now heavily dependent on desalination plants, which in themselves exacerbate the water imbalance. As you can see, the less water there is in the ground, the drier and therefore hotter the land becomes.

These places of water stress epitomize forgetting the language of water. There is no longer a functioning water cycle; a severely disrupted water balance has led to aquifer depletion and the land drying out.

The founding of The Flow Partnership in 2011 was prompted by witnessing these alarming trends in the water balance of the planet and the changes communities were capable of making to address that imbalance. We started with colleagues and friends from India and eventually gained partners globally to make it fashionable to speak this language again. We were first inspired by the work of Rajendra Singh[1] and his village communities who had revived seven rivers in arid Rajasthan,

India. We wanted to start an organization composed of a network of collaborative partnerships that would allow people to understand their role in keeping the local water cycle healthy and by putting a spotlight on questions like:

- What is the local language of water? Who knows it and how can we share this knowledge between local communities?
- What are communities doing successfully in managing their water locally?
- What action on the ground can be taken by individuals to contribute to a functioning water cycle?
- How can this local water-balancing work impact global climate imbalances for the better?

As we delved further into the successful work of village communities in averting droughts and floods, we started hearing countless stories of successful water-retention work done by communities from across the world. We were really surprised! Why wasn't this work getting global attention if it was as successful as, for example, the case of Rajendra Singh who, along with his communities and village "engineers," regenerated seven dead rivers? And why was this work not spreading among global communities so all those facing water stress would have access to the knowledge they would need and the actions they could take?

Quickly we realized that community voices were being heard only through research papers and a few well-meaning educated folk. The stories of practitioners on the ground remained largely untold; sustainability conferences, policy reports, research projects, and academic authorities left little room for the unfiltered voices of those conducting the actual work, speaking through the lenses of their own identities. Communities did not seem to have a global voice of their own, nor was there any route for them to be able to share their successful methods and practices with other communities around the globe. This led us to realize that what needed to happen was a decentralized, ground-level, global movement to bring local water cycles back to health via the learning and actions of these communities managing their water successfully— a people's school to communicate water's simple message. And to get it going, we needed to tell the story of water and demystify the nuances of its language universally. If we told the story well, we thought, some of these amazing hidden stories of landscape transformation might become better known and have more of an impact and be replicated. Which they certainly were.

But that alone was not enough.

We then realized that successful work was not simply about replicating the steps done elsewhere. The way to transform a dried or degraded landscape was to understand the local reasons for the breakdown within its own cultural and social contexts as well—the cut forest, the dried soil, the depleted groundwater, the farmers who migrated and became laborers in the city, the young people who left their villages behind, and so on. The language of water can transform the simplest set of relations between trees, soil, aquifer, and climate from deprivation to flowing life, as long as these changes are introduced by the communities themselves locally. We needed to open the possibility for local communities to welcome back a vibrant, interconnected life with what they could do at their own level through what they knew.

The water cycle happens through all of us; we are all in it and no one is outside of it.

In 2018, in collaboration with global civil engineering firm ARUP (a truly unique engineering firm with a big vision and an effective community engagement program), we started looking into setting up an online water school for local communities across the world to share their successful methods of groundwater recharge and management. We worked on creating language-neutral tools, using video and imagery, that could spread the lessons of the water cycle in different cultures across the world.[2]

In 2021 our vision[3] for setting up global water schools became further developed with the establishment of Water School Africa (WSA) in collaboration with a group of on-the-ground partners in Zimbabwe, Zambia, Burkino Faso, and Kenya. Water School India soon followed in 2023. The Water Schools' platform provides an online forum for networking as well as inter-community sharing of successful methods and practices for harvesting rainwater and slowing the flow of water in the landscapes. They also serve as hubs for enabling actual action on the ground to hold and manage the water in local landscapes.

FROM APATHY TO AWARENESS

It is not enough to hear a good story well told, or even to simply learn the methods that have succeeded. *Action is critical after the inspiration.* But what spurs us to action?

Each autumn in Dartington, Devon, United Kingdom, where we lived and taught at an ecological college, we noticed that the walking path right over the road from us would lead past some particularly badly worked fields, sloping down to a road. The fields consisted of long, vertical rows cultivating grain for animal feed in very lifeless, poor soil (not an earthworm in sight!), channeling rainfall down the incline. The runoff was compounded by harvesting machinery that severely compacted the wet ground in October, leading to flooding in the lane at the bottom of the slope and making it almost impassable from November to March. When the opportunity presented itself, we, along with the Church of England[4] and the Apricot Center Huxhams Cross Farm,[5] became involved in the restoration of the land. After we plowed the land and planted with green manures, the situation radically changed. The soil soon became porous to the rain and rich in microbial life, reducing runoff. Worms started appearing everywhere, breaking down the soil and keeping it well aerated. *And the flooding in the lane below stopped.*

In Rajasthan, India, where traditional ways of looking after the landscape have been pushed to the side by industrial aims, with trees cut down and water-harvesting features fallen into disrepair, poverty has become an accepted fact of life. When one enters into the despair of such a desertified landscape, the disconnection becomes wired into an acceptance of a self-reinforcing state of aridity. We have seen, when invited to work with such communities, that it can take just a few months to radically transform the situation. In Rajasthan, a traditional earthen structure called a *johad* was built to hold 10,000 cubic meters of monsoon rainwater, securing enough water for the community's needs even if there was no rainfall for the next three years.[6] Crops grew, animals returned, livelihoods were assured, and fewer farmers migrated. *And the desertification stopped.*

When meeting an ecologically struggling landscape characterized by debilitating, costly, and disruptive floods or droughts, on the surface there seem to be no obvious signs indicating a disconnection from the local water cycle. There is just a sense of degradation, disharmony, and impoverishment—a state of affairs that can go unchallenged for years. Everyone gets habituated to the flooded road at the end of the field or the abandoned village where the once-plentiful farming opportunities have now vanished along with the young people. Sometimes it requires an outsider to remind the community of their own capability to dialogue with the land, re-speak the language of water, and take necessary action—to remind communities that if they sow green manures, the soil will come alive with the ability to hold water. If the community constructs a huge pond to hold the monsoon water, crops can be watered even during the dry season. The language of water is spoken in that unique story of interconnectedness that is waiting to be discovered in the simple realignment of practices that return the land, soil, and climate to a vibrant state.

A FORGOTTEN LANGUAGE: SHIFTING SANDS DOWN THE AGES

How did we get to this point of forgetting and ignoring the language of water and its inextricable connection with our lives?

Starting in the 1850s, there was a change in the mindset of industrial societies that emancipated humanity from its relation to the land. Old maps show winding rivers being straightened. The zigzag meander of the river finding its natural way through the valley was replaced by straightening it so that goods and people could be moved as quickly and directly as possible from point A to point B. Wetland areas were claimed for farming by draining the land. Moving water from the landscapes into towns was handled through plumbing systems. Big reservoirs were built and metal pipes dug underground to deliver water to taps inside the houses, into irrigation outlets in fields, or into industrial plants— which in itself was useful, of course—but the emphasis had shifted from seeing water as a cycle to be maintained by good practices in the context of the environment to viewing water as a commodity to be collected, piped from place to place, and used limitlessly for a life of ease and plenty, consequence-free.

Traditional farming methods first involved speaking the language of water and then seeing how the land could be optimally expressed within it. Over time, the new mastery of engineering overwrote this language.

Civilizations spoke their might through their interpretation of the language of water.

The colonial expansion of European countries around the world brought a style of farming that began by draining the land. Water was collected centrally in reservoirs and then supplied as irrigation to the fields. This practice both took populations away from their traditional knowledge of land management and established the industrialization of crops with high yields that could be exported widely for monetary gain. In time, these practices were augmented with the use of artificial fertilizers in the soil to increase production through formulaic growing without care for the land underneath.

Traditional practices of working with water locally were replaced by systems of centralized supply. Water was imagined as a free resource engineered to meet human needs—pumped into houses, fields, or factories and pumped out as waste—without questioning the resulting pollution of the natural waterways. Laws were passed that allowed sewage to be dumped into rivers—once considered sacred places—a practice that soon became accepted on a global scale. Mixing dirty water with clean, pure drinking water and then spending millions to treat it and millions more on the multiple consequences of ecological degradation—surely somebody could see the folly of that. But it seems not.

Places like Barichara in Colombia, which had plentiful water in 72 creeks running through the town, changed to an aqueduct system pumping water in from outside, while their local creeks dried up or became filthy, unusable sewers. The forest was cut down, exacerbating the dry conditions. Water shortages and a steadily decreasing water table were the result. This is one example out of a few million across the whole planet now following this same formula. The language of water, when reduced to an engineering challenge, loses its meaningful relationship to the environment that is at the source of its transformational potential. The consequences of depleted soils, total dependence of the farmers on expensive chemicals and fertilizers, and lowered water tables have only now gradually started to become apparent. The chaos brought about by unsupportable temperatures and extreme storms is the sum expression of all our separate actions that have treated water as a limitless, easily accessible commodity external to us and needing no love and care to exist.

RESTORING THE WATER BALANCE

Most people no longer consciously understand the critical role of holding the water that falls from the sky. How many landscapes across the world are drying up or flooding as a result? In the United Kingdom, where we live, the climate is also undergoing change. The United Kingdom is now beset by summer drought when historically it was bemoaned for *always* raining. The storms have increased and the rain falls in more intense bursts, leading to exacerbated winter flooding.

Floods and droughts are both disruptions caused by mismanaging the water cycle. They are opposite sides of the same coin and the strategies and interventions to manage each can often be similar. Bringing back that understanding of a healthy water cycle by retaining water in

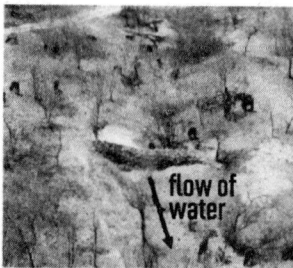

Staggered leaky dams for holding water to relieve a drought Muonde/Zimbabwe

flow of water

Staggered leaky dams for holding water to relieve a flood UK

Leaky dams across the world. Image courtesy of The Flow Partnership.

the landscape is perhaps the single largest action we can collectively take today.

Draining the land for controlled agricultural production may work for a time. An engineered solution to control flooding or droughts can also work up to a point and alleviate severe distress. But it is the multiple small actions, spoken in the language of water by the millions of small communities living on the land across the planet, that decide whether the water cycle speaks to our sustained health or our dramatic decline.

As storms with record rainfalls become more frequent, the engineered wall to keep the river out of the town can only be built so high. In most places, small-scale community solutions can address the problem *before* that high wall needs to be built by the engineers.

In the United Kingdom, growing evidence from successful small-scale trials[7] of nature-based solutions, on tributaries rather than main rivers, led to the government allocating £15 million for further pilot projects across the country.[8] Yet it is still the case that nature-based solutions are seen as having much less impact than engineering manipulations controlling the material environment. Almost no attention is paid to the fact that there are actual projects on the ground that are inexpensive, transformative, and restorative, waiting to make a difference.

Here are some of the common criticisms of this on-the-ground work of communities managing the water themselves:

Critique: Funders, government policymakers, and large organizations do not have the time or skills to handle bringing communities together; they need to take quick action with a set of criteria to test whether the work is done to their satisfaction.

Response: Getting the community on board to restore traditional water practices is key to taking transformative actions that will resolve the problem for the long term. The community should always be at the center of changing their relationship to water.

- In our stories from Colombia, you can see how community awakening can happen through the children.

- In our stories from India, the *Jal Saheli* women decide that they should be the ones who restore the holding of water as they are the ones who fetch and use it in their homes.
- In our stories from Belford in the United Kingdom, it was the farmers meeting in the pub who took the problem of the village flooding (which the authorities could not address) into their own hands.

It is key to involve the community before a project actually begins. Bringing the community on board at the beginning solves countless challenges of ownership, maintenance, and commitment once the engineers have left.

Critique: The water-holding community projects are proven at small scale and are useful additions to larger strategies, but one needs engineering skills to truly address the problem on a macro scale.

Response: This is not an either-or scenario. One needs both the big engineering solution and the soft engineering work of the communities.

We will demonstrate a range of completed projects happening at scale:

- In Dujiangyan, China, in 256 BCE, a hugely creative construction entirely without computers, dynamite, or modeling shaped the river to avoid frequent flooding and provide irrigation to 5,300 square kilometers of farmable land.
- Muonde Trust initiatives in Mazvihwa, Zimbabwe, have systematically addressed the degradation of a landscape through individual farm initiatives.
- In Rajasthan, India, seven rivers and their catchments were revived by Rajendra Singh and local communities.

Critique: The traditional ways of holding water are different all around the world. There is no one method that governments could implement as a global solution that could be rolled out in all countries.

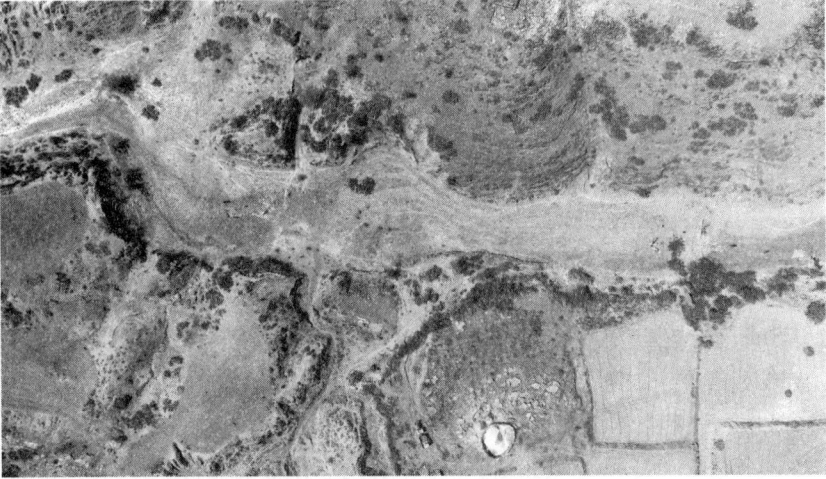

Before: Saseri, Rajasthan, before the work. Image courtesy of The Flow Partnership.

After: The Saseri water bank and its transformation. Image courtesy of The Flow Partnership.

Response: The language of water is universal. If we cut down the forest, dry out the soil, and deplete the aquifer, then the global climate becomes the expression of those actions. If we replant the trees, rehydrate the soil, and recharge the aquifer, the climate begins to express itself beneficially once again.

In this book we will travel to nearly every continent, visiting beaver sites in the United States, taking a creek walk in South America, building rain gardens in Europe, dreaming water in Australia, slowing the flow in the United Kingdom, and planting water in Africa, speaking the universal language of water with the peoples of that land to keep our planet's climate in balance.

Qatar in the Middle East, Burkina Faso in Africa, and Rajasthan in India are at equivalent latitudes. These communities face similar challenges of aridity and drought. Previous eras of forestation and seasonal rainwater recharged the aquifer and have left behind accessible groundwater. The different regions have developed comparable techniques to look after this groundwater, replenishing it with the seasonal rains and carefully balancing what water is taken out for human needs.

Zephaniah Phiri, a much loved water warrior from the Muonde Trust in Zimbabwe, often likened holding water in ponds to creating an immigration chamber for water: "This is where the water comes in before it goes anywhere else . . . and look at the transformation in the land from holding this water."

In addressing how water is harvested within different traditions, we recognize the collective benefit that the water cycle has on the planet. Speaking the shared language of water helps us understand the future of the world as our common concern.

TRANSLATING THOUGHT INTO ACTION

Globally, both in rural and in urban areas, a realization is dawning that we need to change the current system of infinite extraction, exploitation, and pollution of the earth's resources, especially water.

The language of water can still be found spoken in the traditional memories and actions of communities around the world. A particularly severe cyclone, or the breakdown of community farming because of drought, or repeated floods gushing through a town awaken communities to the fact that we cannot control nature. It reawakens the memories of the wisdom of their ancestors.

The chapters in this book present the stories or techniques of a community in one stage or another, following the universal alphabet of the language of water to return a deteriorated landscape back to health. These communities have come together at points of crisis to take care of water within the land, discovering in their traditions how to hold water in the landscape so that the storms and rains are not lost in runoff but held for the next dry season's crops.

Local community water sources are like a country's bank—the holder of its wealth. These water banks hold a wealth that is far more valuable than money. Water harvested from the rain benefits local livelihoods, enables soil to regenerate, and creates a functioning, healthy small water cycle, which in turn contributes to managing and mitigating the severe effects of global climate change.

A common picture begins to emerge from these myriad communities working on the land in the different continents. They are all reading the runes in the landscape, holding their water as it falls from the sky, and allowing it to sink into the aquifers: recharging the underground supplies for another day—another generation—and regenerating the landscape around them.

The prospect of mitigating the worst effects of global climate change through community-led local measures that follow the universal principle of holding water to regenerate landscapes is immense. These measures hold the potential to alleviate dwindling freshwater supplies on the planet and enable tangible, exponential climate change action on the ground. The work and impressions of the communities presented in these chapters give a flavor of what is possible when the regeneration of the planet is held in the hands of those who live on the land and care for its water.

The Flow Partnership's work in India, Africa, Europe, and other places where there is drought or flooding demonstrates that reversing a severe drought or preventing future drought conditions is often as simple as creating locally relevant water-holding structures at strategic points in the landscape. Often these methods of holding water were what our ancestors did but have now fallen out of fashion with modern life and its intellectual solutions based on the myth of infinite growth and perpetual availability of resources. These low-cost, simple structures can be constructed by communities themselves, enabling the people to become water-rich and self-sufficient far into the future.

The stories from these communities show us again and again how the water balance of the planet can return. We hear this common language of water being spoken across the planet, across millennia.

When this language has been forgotten or disregarded, civilizations have vanished, but when it is remembered, the age has thrived and prospered. Through collaborating with communities, we can relearn to speak the language of water and add dimension to the discourses of science and technology. For every story and every heroine or hero we applaud in the following pages, there are countless more, and we salute all of them even if it is impossible to mention them all by name. The movement for a community-driven, decentralized movement that teaches the language of water and how to hold it within our lands is alive and gaining steam across the planet.

2

Rajasthan, India: A Three-Day PhD in the Language of Water

Rajasthan in India has always been a dry state as far back as anyone can remember. In the deserts of Rajasthan, the relentless sun and miles of sand speak of scant life. Yet this state also has the most populous desert in the world—the Thar desert. How did the nomadic people of the desert live for centuries in that harsh heat with little water?

The Thar desert is like any other desert in the world: sandy, dune-filled, hot, and harsh. The Aravalli mountains, considered one of the oldest mountain ranges in the world, are situated in the northeast, with their unique geology, climate, and ecosystem serving as a break in the spread of desert.

The people of the state of Rajasthan have learned to live with this hot and dry climate, developing extraordinarily simple yet landscape-friendly methods to hold the water that falls from the sky during the monsoon months of July to September every year. In this chapter, we will look at the language of water in the desert, flowing through the mountain ranges on its edges.

In the last few decades, a bright light was shone on these traditional ancient ways of

Anupam Mishra.

reading water in the desert landscape of Rajasthan by a gentle, wise, and soft-spoken man named Anupam Mishra (1948–2016). He spent his life researching the language of water as spoken by small communities and their traditional water-harvesting systems in the vast hinterlands of rural India, with a focus in particular on Rajasthan.

NATURE'S PRINCIPLES UNCOVERED BY ANUPAM MISHRA

Anupam Mishra argued that water security or insecurity is not dependent on nature alone—it depends on nature *and* culture. There could be drought in areas with high rainfall, and plenty of water in areas with low rainfall such as in Rajasthan. His books, *The Radiant Raindrops of Rajasthan* and *The Ponds Are Still Relevant*, which were written more than 25 years ago and bring together his research on the different ancient water-holding techniques of the villagers, have reached hundreds of thousands of readers.[1] Anupam Mishra believed in copyleft, not copyright, and all his work is freely and widely available for anyone to use, learn, read, distribute, or take inspiration from.[2] After all, it was not his work. It was the work of the desert landscapes and the ingenious traditional communities that lived in them.

Annie Montaut's preface to the English translation of Mishra's work says the following:

> As Anupam Mishra explains with so much sensitivity and discernment, the people of Rajasthan did not wait for manna to drop from heaven. Instead, they evolved a whole *riti* (practice) or *voj* (voice) around their *shram* (labor) in the field of water conservation. A practice established on a deep partnership between nature (the environment), human action and its ethical as well as religious framework. The same spirit permeates Anupam Mishra's work as well as that of the Gandhi Peace Foundation, the publisher of the original Hindi version of *Rajasthan Ki Rajat Boonden*.[3]

According to the environmental news outlet *Down to Earth*, "A founding member of the Gandhi Peace Foundation, Mishra remained a staunch advocate of decentralized water storage systems like *baolis* (step wells), *kuis* (wells with small diameters), *chaals* (small water body along a slope) and *johads* (tanks that are fed by earthen check dams) that can help communities withstand drought."[4]

The desert in India receives the lowest rainfall in the country: the water is 300 feet deep and most of the water is saline. Rarely does one see clouds in this region of the desert, yet there are 40 different names for clouds in the local dialects spoken there. The implication is obvious—water has a greater meaning for these desert folk; the water carriers—the clouds—are awaited eagerly, as their appearance signifies the continuance of their life in the desert.

This is what Mishra said about the different kinds of water structures in Rajasthan that hold what the clouds brought with them, which of course pertains to our reading of the language of water:

> Keep in mind slopes and catchment and tankas and then the water will go where it should go, along the slope, through the catchment into the *tanka* (tank). Collecting 100,000 liters in one *tanka*, which is pure, sweet, clean drinking water is the result. Rainwater harvesting is not just another project in the lives of desert people—it is their life itself that depends on it.
>
> Put stepwells[5] and look at how they were built for desert people to access the water in the well as the level dropped, and in the meantime culture, and religion happened on top in the beautiful structures housing that well.
>
> Combine engineering with aesthetics and art. The villagers used to paint or sculpt local animals on the water bodies and as they filled up, the level on the animals would tell them how many months of water they had.[6]

Anupam said that making, planning, and maintaining your water bodies was part of the work of life itself. And the secret to long-lasting, beautiful, functioning ponds starts with a respect for common property as if it is your own.

Above ground in the Thar desert, the sand was dry and held little water, yet the villagers knew where a belt of gypsum between the sand and the underground aquifers of saline water allowed sweet and cool water to run. They constructed wells (called *beris* or *kuis*) along this belt of gypsum that provided them with the water they needed to drink, which became the lifelines of the Thar desert. Villages sprung up in these areas. This wide belt of gypsum kept the rainwater from percolating into the salty groundwater below while the sand above covered it and stopped the sweet water from evaporating in the strong sun.[6]

How did the villagers know where this belt of gypsum and its sweet cool water ran? Now that was an art, understood through the experience of old, wise eyes. Where the sand showed moisture underneath, there was sure to be water. Unlike clay, which becomes lumpy and cracked when dry, sand remains the same on top. The moisture in sand simply goes down a layer and stays there. Experienced eyes could spot the nearly imperceptible difference in the color of sand that has moisture beneath as opposed to sand that has no water underneath.

The moisture that lived a little below the surface was waiting for an experienced reader to notice the difference and dig a small well to reach that reservoir of sweet water flowing along the gypsum belt. Knowing, reading, and acting on the knowledge of the language of water truly meant the difference between life or certain death. This belt of freshwater enabled the Thar desert to be the most populous desert in the world.

In other parts of Rajasthan, while wells accessed sweet, cool water from the underground aquifers not that far below the surface, the groundwater was replenished each year by structures they built upstream to hold every drop that fell from the sky. It was collected upstream by building interventions into the landscape where the maximum amount of water would pool and sink into the ground, recharging the aquifers and allowing the water in the wells downstream to rise. These long earthen embankments, or bunds, were often community-owned and built around the edges of farmlands to arrest and hold the water at the bottom of the slope. In the Aravali catchments, these interventions were often in the shape of a crescent and were called *johads*. Anupam Mishra succeeded in capturing the imagination, romance, and respect for those water-literate desert people whose wisdom had enriched their lives for millennia with little need of any sophisticated modern science.

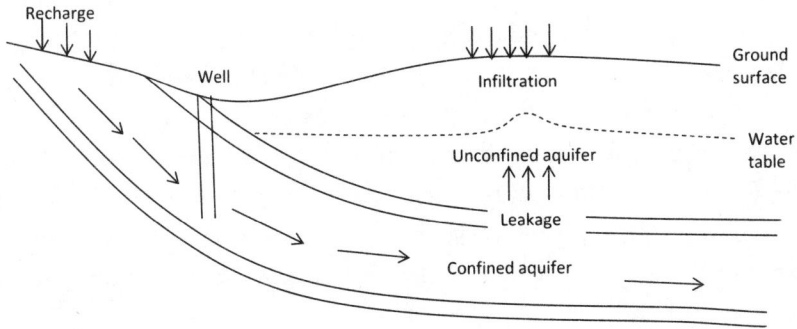

The system of recharge into the aquifer.

PUTTING PRINCIPLES INTO PRACTICE

These traditional ways of working with the water fell out of favor when European colonists came, promising water in a tap and ignoring that these traditional water-storage structures were the source of the water that would run in those taps. The taps, however, did not come, and soon the water stopped coming as well, with the region sliding into aridity and degradation. In the 1980s, the remarkable Rajendra Singh began to revive, transform, and scale up these traditional water structures over huge areas.

We first met Rajendra Singh when we were asked to invite a water hero to speak along with us at a water conference in Portugal in 2012. Having long wanted to meet him, we suggested he should be invited, and we got along famously upon our first meeting. We went on to visit his work in India and saw firsthand the impressive work his villagers had done across the seven catchments and rivers they had revived. From then on, we worked together to create greater international visibility to this approach and to enable a network of similar community-driven water-retention work across the world. From this joint work, we set up The Flow Partnership as a registered charity in the United Kingdom.

Rajendra Singh.

Rajendra, affectionately called the Waterman of India, had two great mentors in Mohandas Gandhi and Anupam Mishra. He trained as an Ayurvedic doctor living in the city of Jaipur and was often engaged in activism for social causes such as eradicating alcoholism and treating the sick with Ayurvedic medicine. One day, when he was about 26 years old, Rajendra and three of his friends, inspired by the ideals of Gandhi, had the thought to go and serve the poor in the villages by taking medicines to them and providing them with basic education so they could learn to read and write.

They sold many of their possessions, said goodbye to their families, and boarded a bus for the hinterland of Rajasthan. When the bus conductor sought to sell them a ticket and asked for the destination, they said "last stop!" Over 24 hours later, they landed in the village of Gopalpura in Kishori, in the Alwar district, then a remote village in Rajasthan in the Aravalli hills. It was dusk by the time the bus stopped and let them out so they couldn't see much of the area where they had disembarked and what this last stop could hold for them. The villagers gathered, a bit fearful to see these bearded, strong-looking young men get off the bus. "Were they terrorists?" they wondered to themselves. But as is the custom in most Indian households, the villagers were hospitable to the young men and directed them to the temple courtyard to sleep. They even gave the youngsters a bite to eat before retiring for the night.

The next day, Rajendra and his three friends met with the villagers and told them of their plans to treat those who were ill and set up a school and educate them, zealously beginning to fulfill their noble intentions in earnest. The village was not only remote but dry, dusty, and with nothing growing but a few thorny scrub bushes and a few emaciated goats listlessly lying around in the extreme heat. In their minds, this was exactly the kind of place they had hoped to give their free medicine and education in order to help the villagers escape their poverty and aspire toward a decent life.

Rajendra set up a place in the village square where he could treat any ill villagers who might want to seek help. The others walked the length and breadth of the village, knocking on doors asking if the children were going to school and, if not, whether they could help them learn to read and write and make something of themselves. This went on for a few days. Very soon the visitors realized that, while the villagers were kind and polite to them, almost humoring them, no one was taking up their offers of free medicine and education!

Rajendra's three friends finally threw up their hands in disgust, saying that the villagers were hopeless and that no one could do anything for them if they didn't want what was being given to them for free. They left the village to go back to the city. Rajendra, however, kept feeling that he was not looking carefully enough and was missing something. On a walk through the village, he came upon an old man sitting on a rock, watching his approach intently. The old man, affectionately called Mangu Kaka (uncle), patted him on the head and told him he was a good young man with good intentions and the villagers were grateful that Rajendra wanted to do something for them. He then asked Rajendra, "Have you seen the state of the village? Can you not see that we don't need medicine or education, both of which we could purchase if we earned enough money? What we actually need is water. Why don't you give us water instead?" Rajendra replied, "What do you mean, give you water? I'm not an engineer. I'm a doctor. I don't know how to give you water."

The old man then said that Rajendra didn't have to be an engineer to give them water. "I can teach you how to give us water. Do you want me to teach you how to do that?"

This surprised Rajendra. If the old man knew how to get the water, then why were they in this pathetic drought state that kept them in misery and poverty? His curiosity piqued, Rajendra said yes and the old man told him to come back the next day with a spade so he could show him where and how to get the water.

The next day Rajendra presented a funny sight. Somehow he had borrowed a spade and, carrying it on his shoulder, made his way down the dusty village street with everyone in the village who was watching him thinking, "He's a city lad, not used to hard work. He'll run away tomorrow." Running after him at a safe distance behind was a giggling band of children, teasing him and asking him what the spade was for, all the while following him as he walked on to meet Mangu Kaka. Sure enough, there was Mangu Kaka at the designated spot, who gave a small hint of a smile before pointing to the ground and ordering him to start digging. "Don't ask questions. Just dig. Let me see what you are made of."

Rajendra dug the whole day. And the next and then the next and then the next. By the fifth day, the village realized he was serious. Soon others in the village joined him in the digging. When finished, they had a hole in the ground that Mangu Kaka was calling a *johad* (pond)! All the while, during the tiring work of hand digging, they wondered why they were being asked to dig here and what the *johad* might do. Why did Mangu Kaka pick this precise spot in the landscape? Rajendra asked Mangu Kaka these questions; what ensued was what Rajendra calls his PhD in the language of water in just three days.

THE LANGUAGE OF WATER AS TAUGHT BY MANGU KAKA

Mangu Kaka took Rajendra on a walk about the region and showed him the lay of the land and the dry wells that lacked water. At times, he put Rajendra in a large leather basket and lowered him 150 feet deep into a well. (No worries about drowning—there was no water in the well, for they were all dry!) Rajendra lost count but thinks he must have gone up and down at least 25 wells.

Mangu lowered him into different kinds of wells, which showed him the geological layers within the landscape. Each time he would come back up, he was inevitably quizzed: *what did he see in the well?* In some places he noticed horizontal fractures and in others he noticed vertical fractures. Where there were vertical fractures, the trees were tall with long roots going deep, making the trees strong. Where there were horizontal fractures, there were no trees, just small shrubs. The vertical fractures allowed the water to percolate deep and the long roots of the trees sought the water out.

In this way, Mangu taught Rajendra the language of water in the underground aquifers through trees, existing vegetation, and the geology on the surface.

Villager from Mangu Kaka's area: Illustration Dilip Chinchalkar from *The Ponds Are Still Relevant.*

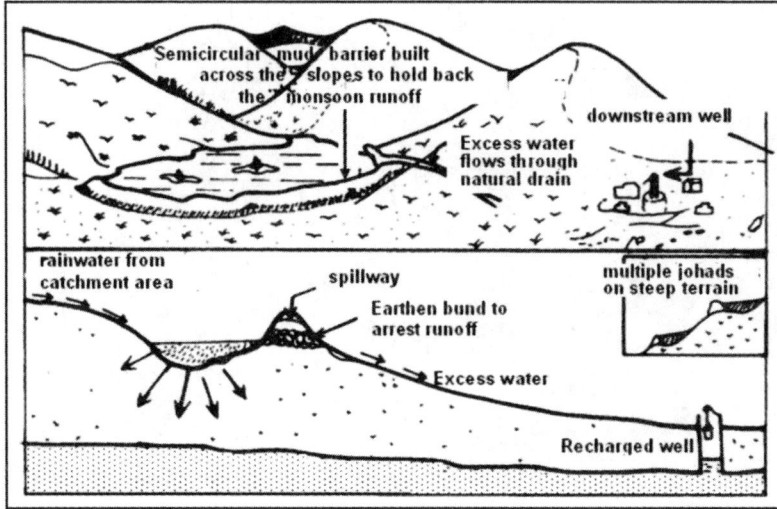

Johad cross-section: courtesy TBS, Rajasthan

The next lesson was how to select a particular spot to site a water pond. The answer? Find the point where the water naturally collects before running off and becoming a moving stream.

This was village wisdom. It had kept them in good stead over centuries. For millennia, the desert communities of Rajasthan had been building their *johads* in such places in the landscape to hold every drop of water that fell from the sky in the monsoon months and stop it from running away. Holding the rainwater upstream recharged the underground aquifers and the downstream wells remained full of water year-round.

From that first *johad* that Rajendra built with Mangu Kaka and the villagers of Gopalpura, which collected water in the very first monsoon and recharged the downstream wells, they went on to build 500 such structures in that catchment alone, reviving their dead Arvari River to flow seasonally after 4 years and perennially after 12.

Over the following 28 years, Rajendra and his teams of villagers went on to build 12,000 such structures, of every size, sometimes called *johads*, sometimes called *medhbandis*, and sometimes called *paals* depending on the kind of structure suitable to the particular landscape. These 12,000 structures were strategically situated to work with the geology to rejuvenate the water table. While many policymakers still see traditional

Community building a johad:
Illustration Dilip Chinchalkar
from *The Ponds Are Still Relevant.*

natural catchment management as a small-scale add-on to engineered structures, Rajendra Singh had the foresight and determination to show how this work can be scaled up to strategically regenerate entire catchments. Seven rivers were revived, which are flowing and providing ample water to the villages along their banks even to this day.

In 2015, Rajendra Singh received the Stockholm Water Prize for this work, which he accepted on behalf of all those villages and communities with whom he had worked to revive their rivers and lives.

Let's return to the question of why, if they already knew what to do, the villagers didn't do this restoration themselves before Rajendra came. The answer to this question was complex and simple at the same time. Extreme poverty and neglect had caused all the youngsters of the village to migrate to the cities in search of work; the old people and parents with children left behind felt apathetic and had lost any hope of a vibrant life. They fell into the habit of bemoaning their lost young ones, the lost water, and the loss of life itself. That was the simple answer. The more complicated answer was cultural. Rival political parties in the village said they would not let the other party do the work. It all disintegrated to the point of extreme neglect, drought, and famine with no one doing any of the work until the day inspiration and energy returned to them in the form of a young man called Rajendra Singh, who rolled up his sleeves and started digging, reminding them that they all spoke the language of water well and enabling them to put differences and apathy aside to become water rich once more.

There's one final twist in the tale that had a far-reaching effect in energizing the villages far and wide in reassuming responsibility for the land and the water cycle. Once the Arvari River started flowing, the villages—all 80 of them along the river—now had an agricultural livelihood as well as fishing from the river to supplement their diets and income. This had been going on for a few months when one day the village head of Gopalpura received an official-looking document from the government—that same government they hadn't heard from in years while they were starving and dying due to lack of water. The letter said something like, "We have noticed that there is a river flowing in your area with fish in it. Please be aware that to fish from that river you need to apply for a license from the government."

It is not hard to imagine the shock, outrage, and disgust that rippled through all the villages on hearing this. A local meeting was called for all the villagers from these 80 villages along the banks of the Arvari River, and the very first River Parliament was born. For a few days, they discussed the state of the villages before and after the return of the river, their responsibilities to keep it flowing, and what rules they needed to follow to ensure they never returned to that state of drought and poverty again.

River Parliament. Image courtesy of TBS.

And then they wrote their reply to the government. They said, "Yes, you are right, there is now a river flowing in our region. Yes, you are right that there are now fish in said river. Yes, you are right that a license should be taken to fish in the river. However, the one thing you are *not* right about is the notion that we should have to apply to you for the licence to fish in this river. It is *you* who need *our* permission to fish here. The river belongs to us. We revived the river with our common, collective hard work. The Arvari River Parliament may have no legal authority but we have a moral authority derived from the fact that we gave life to this river again."

After some back and forth, the government finally agreed that the river belonged to the 80 villages along its banks and promised to uphold their moral authority in making decisions for the distribution of its waters and its fish. The River Parliament has two representatives from each of the villages along the river. They convene twice a year to safeguard the efforts of the communities along the river watershed by developing a coherent water use policy of the revived river, enforcing rules to govern the integrated management for interlinked natural resources like water, soils, and the forest for the well-being of the villagers and other life forms. It also functions as the intermediary in village-level conflicts, resolving them through cordial discussion and mutual consent. The River Parliament follows the Gandhian ethos of a participatory, equitable, and decentralized paradigm for water management (*Jal Swaraj*), where decisions are made at the grassroots level and not by centralized institutions—an unparalleled victory for the community and community-driven decentralized water management.

NINE SIMPLE YET PROFOUND LESSONS RAJENDRA SINGH LEARNED ON HIS JOURNEY WITH WATER REVIVAL

In his own words:

1. When I completed digging the first water body, and it filled up with the monsoon water and I saw the water in the downstream wells, I understood water and how it moves in the landscape. That was what I needed to know to build more *johads*. Water conser-

vation stops soil erosion and that stops the drought. When water came back, more vegetation came back and then more rain and that cooled the land as well.

2. Water is best stored underground to reduce evaporation loss. To keep the water and earth safe from the sun, increase recharge and decrease discharge, increase the deposit in the underground water bank (aquifer).

3. When underground aquifers recharge, the vegetation returns and ample water starts to feed the small rivulets that flow in the landscape.

4. When the earth is hot, the clouds go away. When the earth is green, the vegetation starts that water cycle again, cooling the earth through the evaporation of the water in its natural cycle. When the earth is cool, it attracts the clouds for more rain—science with common sense.

5. Where there are steep slopes, you create a convex design check dam using local lime and sand. Where there is lime content in the soil, you make a concave structure.

6. The education system today teaches you the way to control floods and water. Nature is not to be controlled, it is to be respected, loved, and worked with in collaboration. That's a key lesson in the language of water.

7. First I learned from the community and the elders, and then I trained the young people. It's a community effort. We think of the common future of both community and nature. When we can flow with nature, nature gives us the flow of life and peace.

8. If you destroy water, you destroy life. Birds, ants, elephants, and humans all have the same equal right to water. Water is the life of the whole planet, and corporatization of it is only for a few.

9. We need a communitization of water—it's a common property. Local people, small tools, and small ponds holding water are the future.

CASE STUDY: ARVARI RIVER WATERSHED (ALWAR, RAJASTHAN, INDIA)[7]

Johad water retention in the landscape: courtesy R. Varma

Description

Rajasthan, which is India's largest state by area, comprises 10.4 percent of India's land area, 60 percent of which is arid and 40 percent semiarid, with 90 percent of annual rainfall occurring in the monsoon months from July to September. Agriculture in Rajasthan relies heavily on groundwater, and overexploitation of this resource has led to frequent droughts and an exodus from the villages to the cities.

The headwaters of the 90-kilometer-long Arvari River rise from the Aravalli hills on a geology of folded and fractured conglomerate, grit, and arkosic quartzite rock through which groundwater flows. The Arvari watershed (with an area of 476 square kilometers) is semiarid, comprising 46 micro-watersheds with a covering sand sheet on flatter land downstream that protects the aquifer, also comprising soils that are fertile but which lack water.[8,9]

In the 1980s Rajasthan suffered several years of drought, leaving remote villages with hardly any water to meet the needs of the inhabitants. Removal of water management responsibilities from local control led to disengagement, a shift in perception of the value of water, and a lack of self-responsibility of local people from the management of these natural assets. This led to widespread abandonment and the degradation of community water-management structures.[10]

The model of engagement used by Tarun Bharat Sangh (TBS), an NGO doing water retention with communities in India, to revive the Arvari River ensured that every household in a village located along the river watershed that was suffering from drought would contribute something according to capacity—either land, material, or labor. TBS would raise the necessary 30 to 70 percent of the cash required for the building work. The work would be done professionally or by the villagers themselves depending on the availability of the funds. The first few *johads* were built in strategic places in the landscape so as to recharge the underground aquifer as well as the surplus to overflow on the path of the river. These were self-funded by the richer villagers in the area. Between 1985 and 1995 some 350 *johads* were built on the Arvari River watershed with the help of funds from various European development agencies working in partnership with TBS.

Impacts of the Project

As the *johads* got built, the river started to flow seasonally, and within 12 years the Arvari river became perennial. Rajendra Singh often says, "We never realized that we were recharging a river. Our effort was just to catch and allow water to percolate underground."

"Work done this way fulfills the need of self-reliance of local people," says Singh. "In a small project everyone participates in the decision making. And everyone contributes something to the whole process. In this process, each and every one becomes invested in the successful outcomes. The community gets employment with an overwhelming feeling of ownership and being in control of their destiny brought about by their efforts."

Before the *johads* were built:

- 35 percent of the rainwater was lost immediately as seasonal runoff.
- 50 percent was lost due to evaporation or transpiration.
- Only 15 percent of the rainfall naturally recharged the groundwater.

After the *johads* were built:

- Groundwater was recharged by an additional 20 percent, raising the water table.
- Rainfall seepage into the river increased by 17 percent in non-monsoon months, reviving the Arvari River and making it perennial.
- Seasonal direct runoff of the rainwater has now come down from over 35 percent to 10 percent.
- There was an increase in soil moisture.
- 22 percent of the runoff (excluding the 10 percent seasonal runoff during the monsoon) is now better regulated and spread out over the year. This has been crucial in reviving these small rivers. If this runoff had not been regulated, the rivulets would not flow throughout the year.[11]

Ecosystem Service Benefits of Project

The Arvari River watershed is still drought free, with ample groundwater available year-round. The watershed has experienced a dramatic increase in vegetation, from no crop growth prior to 1985 to up to three harvests a year since 1997.

Lessons Learned

The River Parliament has framed 11 rules for the river basin conservation and management on issues such as which crops to grow, when the water can be drawn from the wells, how much of it can be drawn, restrictions on cutting down trees along the watershed without the permission of the parliament, and limiting food and grain production to smaller scales so as not to overuse the groundwater.

Most important, the success of this River Parliament has ensured that every member of the rural community is forever engaged in the process; they bear the onus of ownership of the river and its waters responsibly.

COMMUNITY TRADITIONAL WISDOM

Community-driven, decentralized water management using traditional methods of rainwater harvesting (RWH) has led to the rejuvenation of many seasonal rivulets as perennial rivers in this dry and arid part of India. People's participation is a prerequisite to rejuvenating the water cycle. To remind them of the beauty of life and its interconnectedness with water, the desert villagers of Rajasthan painted on their *johads* the following traditional drawing.

Traditional water painting from Sita Bawdi, symbolizing the waterwork.

Commenting upon this artwork, Anupam Mishra noted that **"the center of life is water. These are the beautiful waves and these are the stairs to get to the water and these are the trees and these are the flowers which add fragrance to our lives. This is the message of the desert."**[12]

3

China: Dujiangyan

Dujiangyan (doo-jang-i-yan): A community water irrigation system from 2,250 years ago.

 **Kan**

Water

ANCIENT WATER WISDOM

There are some places where the language of water has never been forgotten. In Dujiangyan, China, we can see the way modern interventions belong to a continuous practice of stewarding water over thousands of years.

In the ancient Chinese language, water was depicted through the trigram seen above. Master Peng, a feng shui specialist, tells us that the trigram represents a central, fast-moving yang line surrounded by two slower moving yin lines. This one simple character encompasses a whole living process. The inner central line is the sand and stones flowing in the middle of the river, some of which naturally tend to accumulate at the riverbanks (represented by the outer broken lines). Seeing this, we understand immediately that water and soil have a relationship and impact each other.

When first catching sight of Dujiangyan, the 2,250-year-old water irrigation system, it is easy to think that the ancient Chinese must have possessed great computing power to work out these principles of water and apply them over such a huge scale. Not so. There were no computers then. Yet all the water flowing down from the Himalayas on the Chinese side has been tamed into an irrigation system that has worked successfully for thousands of years. To know the language of water is to appreciate the nuances of the way water speaks in its different states. The water cycle naturally transforms water from a raging torrent, to a fast flowing river, to an irrigating canal, to rice-supporting soils, to evapotranspiration into vapor, to clouds, and back to rain over the mountains. The language of water is built up over generations of communities, scholars, and practitioners seeing how these different states of water contain the secret clues to how water exists and moves in the landscape. The language of water to the ancients was not outside water itself, as is told to us in the chemical composition of hydrogen and oxygen. The language of water was deciphered from the qualities that water could take on, whether in rage or in stillness.

This was the language the ancient Chinese knew all too well. Thus in Dujiangyan, all the different properties of water—its ability to carry silt to one place where it could be excavated, its ability to irrigate the soil, and its ability to be carried over long distances—were naturally included as part of the system's design. Water to the ancients was elemental, with a language that had to be understood from the practical knowledge of the properties in its different states. Through this knowledge, the Chinese became experts in working with water at huge scales.

An 18th-century abbot, Wáng Lái-tōng (1723–1799), reported in his main work, "Construction of Dams according to the Heavenly Times and the Earthly Munificence":

I, the unworthy, who have studied the passage of the sun and the moon in the sky, have known their numbers, have mastered the irrigation work, also possess the maps and tables. As a result, the opening of the dams and the dredging will be done in accordance with the transformations of the Heaven and the Earth, in harmony with the ebb and flow of the water, in sync with the seasons and the weather. What is deep will be dredged and what is low will be built up as needed. Then the water will fill the irrigation ditches, the peasants will be peaceful and joyful. Do my insignificant opinions constitute the proper knowledge of the dams and waterworks or no?[1]

The language of water is not theoretical. All the possibilities for water's behavior, from turbulent floods to calm ponds, are included in its story.

THE IRRIGATION SYSTEM

The Dujiangyan flood prevention and irrigation system was constructed around 256 BCE on the Min River, the longest tributary of the Yangtze River, in southwest China (between Chengdu, the Sichuan Basin, and the Tibetan Plateau). The intervention was placed at a critical point where the melting snows and precipitation running off the surrounding mountains slowed down upon reaching the plain.

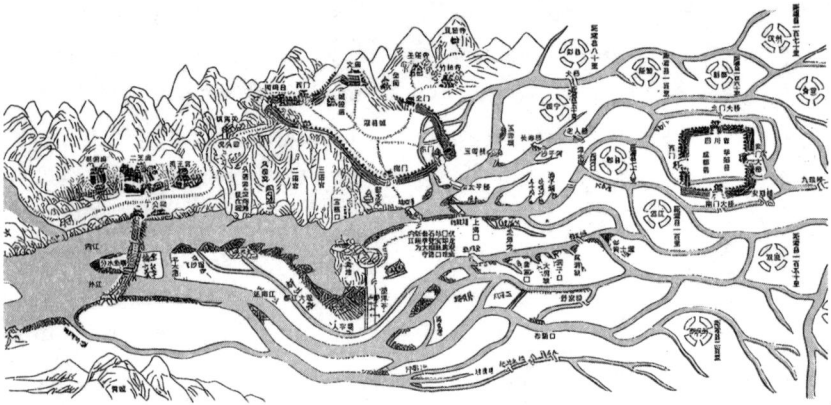

Main view of the site. Image from Annals of Dujiangyan.

Slowing the water caused the silt to separate out, accumulate, and impede further flow. This unruly confluence of forces caused the area to flood frequently. Reading the nature of water in their region, residents devised an ingenious method to channel and divide the water pouring down from the nearby Himalayas. It allowed the silt to collect at a single location, from where it could be removed so the water could run free instead of backing up and flooding. This system provided multiple benefits of flood control, irrigation, water transport, and general water consumption to the local population on the plains.

The water management system consists of three main constructions that work in harmony with one another to prevent flooding and keep the fields well supplied with water.

Wáng Lái-tōng describes the system as follows:

This dam project connects the southern river with the northern reservoirs, it involves dredging the depths and building up the banks, it has lasted over the Qín, Hàn, Jìn, Suí, Táng, Sòng, Yuán, Míng dynasties, till the present dynasty, all in all more than 2,000 years, with the most important element of the system being the levee.

The behavior of the water in this system varies over the year, with the level of water being high, middle, or just reaching the levee, depending on the nature of the waters accumulated high in the mountains, and then flowing down into the river at varying intensity, all gathering in a stretch of one li [half a kilometer], not being able to surpass the levee. The most important requirement is to calculate the parameter of the floods, their wedge-like shapes, their duration and height, and to conduct dredging and building in accordance with the irrigation maps showing the usage of water in the peasants' fields.

The Baoping Kou, or Bottle-Neck Channel, serves as a throat, through which passes water for the 10 neighboring counties, similar to the human windpipe which passes inhalations and exhalations, or to the human blood vessels. If the water flows according to the agricultural maps and timetables, it is not only similar to the blood and air supply in the human body, but also provides for the luxurious growth of the seeds, and ensures the bountiful harvest.

Also, due to this single dam, the waters are divided into millions of irrigation ditches, irrigating all the paddy fields in the 10 counties in the entire province, nourishing the lives of the millions of families, truly being the root of their existence, their savior source; even the smallest error in its work can cause hardship to the people!

Therefore, if we are able to exhaust our hearts in servicing this dam, comprehend the water's behavior, penetrate its transformations—then all the sentient beings will benefit and rejoice unendingly, and our merit will join that of the builder Li-bing himself, from generation to generation![2]

The relation of these features to the landscape and to the river are shown in the following map.

ELEMENTS OF THE DESIGN

The combination of Fish Mouth Levee, Flying Sand Weir, and Bottle-Neck Channel, whose exotic names describe their unique and original design and construction, channel the raging waters from the mountains and irrigate over 5,000 square kilometers of agricultural land in the dry Chengdu plain. Our first task is to describe these features, with the help of abbot Wáng Lái-tōng from 300 years ago.

View of the irrigation system. Image from Sichuan Provincial Water Resources Department

THE YUZUI OR FISH MOUTH LEVEE (鱼嘴)

The first challenge for flow management was how to slow the water as it rushed down uncontrollably from the slopes of the nearby mountains. It was impossible to put any impediment where the water rushed at full momentum, as this would immediately have caused the water to back up and create a flood. Instead, an ingenious feature was designed to use the slope of the land to divide the water into two courses: a direct, deep course that would keep the direction and pace of the outer river and an inner, slower course that would be shallower and curl further around. The structure that divided the waters in this way was created by using bamboo cages filled with stones, gathered from nearby.

The fish-mouth embankments with bamboo cages, intended to divert water, made in the shape of a "fish mouth," must be higher than the surface of the water by 5 chi [1.65 meters], at the initial point of diversion, and must gradually increase in height to 1 zhan [3.3 meters] at the meander cutoff, which is the standard height ratio. —Wáng Lái-tōng

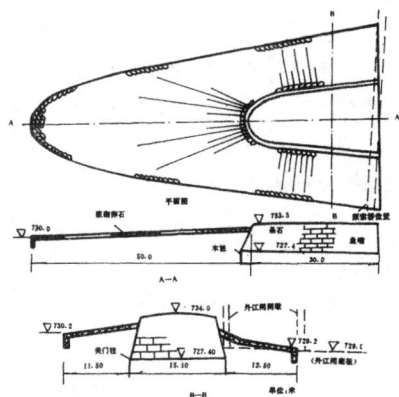

The Fish-Mouth Levee from Annals Of Dujiangyan

图 2-7 1936~1974 年鱼嘴结构图

Another realization was with regard to silt. The slowed-down water in the shallower channel would release the silt conveniently at the bend of the bank, where it could be cleared from this single location. Every year a community ceremony still takes place during which the silt is removed from the place where it tends to accumulate, followed by a festival and celebration.

An old picture of the community festival celebrating the desilting of the banks. Image from News Report of Qingming Water-Releasing Festival.

As to the gravel and stones accumulated in the shallows during the summer and the autumn, clearing every 1 chi [0.33 meter] of them will result in addition of 1 chi of water. Such deepening and dredging must be done until the Beginning of Spring, in order to produce 5 chi [1.65 meter] depth of water. —Wáng Lái-tōng

THE FEISHAYAN OR FLYING SAND WEIR (飞沙堰)

The second intervention that ensured the system was not overwhelmed by the water was constructed at the end of the slow inner detour, near where the two streams reconnected. The purpose of the Flying Sand Weir was to introduce a limit to how much water would be diverted into the irrigation system. When the water flow was high, the excess water would flow over the Flying Sand Weir and rejoin the outer flow without flooding the channel to the irrigation system.

The weir length must be 100 zhan [330 meters]. It has to be high enough to accumulate spring and summer waters in the reservoir; it also has to be low enough to allow the wild floods of summer and autumn to flow over it into the Yangtze River. This height is called the Minimal Height of a Dam Rule. —Wáng Lái-tōng

The turbulence of the water by the weir also further helped to drain silt and sediment from the river. Initially this structure was built with bamboo cages and stones, but in modern times this has now been replaced by a concrete weir.

The ingenuity of this structure limits the rush of the river between the Fish Mouth and the weir so that a known and predictable flow of water can then be harnessed into the irrigation system without controlling the whole river between high concrete banks. Two structures working together could create a workable flow of water between them that could then be safely channelled into creative use downstream.

THE BAOPING KOU OR BOTTLE-NECK CHANNEL (宝瓶口)

To bring the water into the farming land of the Cheng Du plain, it was necessary to cut a channel through the mountain. The third structure, the Bottle-Neck Channel, was created with a narrow opening, so that any excess water would flow onto the Flying Sand Weir, preventing flooding. The channel has enabled over 5,000 square kilometers of farmland to be successfully irrigated even today.

Without dynamite, a combination of fire and water was used to heat and cool the hard rock of the mountain, allowing the construction of a 20-meter-wide channel over eight years.[3] Without modeling tools, guiding the water must have required intuition and trust in the skill of those who knew the language of water in the landscape.

> Another important requirement concerns the difference in height between the reservoir entrance (Bǎopíngkǒu, "bottle-neck channel") and the water drop gates on the top of the dam. If the dredging of the stretch between them is not done according to seasons, then the dam will not catch the spring river water. —Wáng Lái-tōng

The bottle-neck channel

All these features worked in combination to balance the huge flows and make them available for the irrigation of the plains for productive farmland.

Dujiangyan reaches out to us from 2,250 years ago, illuminating the potential for natural catchment engineering to realize multiple benefits including flood prevention, widespread irrigation, and transportation of water at a huge scale from a single system. When the language of water is spoken well, it continues to teach its lessons through the ages and centuries hence.

While we have many tools today such as rock-breaking dynamite, computer modeling, and land works mechanization, we often lack the imagination to visualize the water in the way that the inventors of Dujiangyan did. In modern-day solutions to floods and droughts, there is a continuing insistence that the work be done by experts, using an understanding of engineering that is often external to the systems on which it is applied.

What can be lost in our modeling mentality is respect for the nature of water as a whole and how it can benefit us while retaining its essential nature. Nowadays, we might tend to treat water as a pure engineering question, addressable through technology and without this deeper dimension.

The climate crisis, landscape desertification, and dramatic flood events are evidence of the fact that we have stopped seeing water in its whole nature. Our engineering solutions are often fragmentary and costly, built to address a single localized problem. We do not sit down and contemplate the whole nature of water, as the ancient Chinese must have done 2,250 years ago, in order to come to an agreement on how we can best protect ourselves from water's extremes and benefit most from its gifts.

In the natural language of water, the different concerns of economists, industrialists, and ecologists can meet and understand each other. What is defensible from a purely scientific, economic, or agricultural argument alone is indefensible when placed within a language that oversees the choice between life and death, decided by the presence or absence of water.

Repairing the banks. Courtesy of Wen Li.

The scale of the solution in Dujiangyan was not beaten or bettered by time, for it is read from the language of the properties of water written in the book of nature itself. The text written in the language of water is still there to this day to be deciphered if we would take the time to read it. The scope of its work is phenomenal, integrating the many imaginative aspects of water in a single work.

> The desirable quality for Qi is harmony:
>
> When the Heavenly qi is harmonious—the rain and dew is ample;
> When the Earthly qi is harmonious—all beings are prosperous;
> When the human qi is harmonious—the intellect is appearing;
> When the water qi is harmonious—the waters are soft and peaceful.
>
> But if the Heavenly qi is not harmonious—there are droughts and floods;
>
> When the Earthly Qi is not harmonious—there are earthquakes and landslides;
> When the human Qi is not harmonious—the sickness starts rising, appearing and disappearing in turns;
> When the water Qi is not harmonious—the waters go sideways, flooding the banks, swelling and rushing. —Wáng Lái-tōng

The purpose of this language is to integrate a system of diverse elements and to communicate the need clearly in order to engage the participation of the community.

> The movement of yin and yang going against each other or in the same direction, always obeys the natural rhythm, as explained above. Therefore, such a model can be used to calculate the construction and repair of dams and the behavior of water in the periods in the different seasons. According to these rules, how should the dam repairs proceed in conjunction with water patterns of behavior each year, after the Beginning of Winter? —Wáng Lái-tōng

The language of water unites the different aspects of water's relation to the landscape in an innovative way and brings the community into relation with the river.

A view of the water in the irrigation system: image from Sichuan Provincial Water Resources Department

In a recent newspaper article,[4] the laid-back, easygoing approach of people from the area was attributed to the irrigation system where all the troubles of life were taken care of and water was made easy to access. The farmers get a plentiful harvest with the water from the river and are left with plenty of free time. The system from 2,250 years ago has changed the lives of the community over millennia and even today. The community is engaged with the collective benefit of the work. They are proudly involved in the upkeep and desilting of the canal system throughout the center of their town.

A local community member putting together the bamboo cages to direct the water to the right channel. Courtesy of Wen Li.

Other work done by the community included making bamboo cages that held stones to channel the water. Creating these stone-filled bamboo cages allowed the annual adjustment of the direction of the water into the channels, to fine-tune the form the water took that year. Thus there was a constant interaction between the community, nature, and the irrigation system.

The language of water is not just a language of technological expertise but a language learned with practice and engagement with water. Anyone who has put their hand in water and watched its eddying passage over a rocky riverbed can speak to the process of slowing and directing its flow.

In fact, the interventions described here were devised in parallel with the building of Taoist monasteries nearby in the Qing Cheng Mountains. In 142 CE the philosopher Zhang Dao-Ling founded his school of Taoism on Mt. Qingcheng, which then expanded as new temples reflected the development of Taoism and the influence of Buddhism. The language of water mirrors aspects of Buddhism and Taoism that hold the individual in balance with the world. Taoism seeks to leave the spirit free, so that its whole nature can be appreciated in its many different expressions. As Master Peng says, to be able to solve the ecological crisis, one must first have the wisdom to know oneself.

Thus the language of water is open to both community participation and to a spiritual story interpreting the behavior of water, joining mainstream theoretical approaches to managing landscapes with communities still in dialogue with the land and the spirit.

Two thousand years ago China had a language of water. Chapter 8 of the Dao de Jing, translated by Yutang Long,[5] describes those qualities of water that also characterize the human spirit.

上善若水。
The best of men is like water;
水善利万物而不争，
Water benefits all things
and does not compete with them.
居众人之所恶，
It dwells in (the lowly) places

that all disdain.

故几于道。

Wherein it comes near to the Tao.

居善地，

In his dwelling,

(the Sage) loves the (lowly) earth;

心善渊，

In his heart,

he loves what is profound;

与善仁，

In his relations with others,

he loves kindness;

言善信，政善治，

In his words, he loves sincerity;

In government, he loves peace;

事善能，动善时。

In business affairs, he loves ability;

In his actions,

He loves choosing the right time.

夫唯不争，故无尤。

It is because he does not contend

that he is without reproach.

4

Language Basics

At the foundation of the language of water is the small water cycle. We all studied it in school. With the decrease in stocks of the freshwater on the planet, the challenge is to make it conscious in our minds once again. The water cycle is connected to every action we take, impacting the health of the climate and the state of the planet. The water cycle refers to the integration of many components: the trees, the vegetation, the flowing water, the rain, the clouds, the aquifers, and the canals that coalesce into a whole characteristic behavior of the climate landscape. The smallest change of a blocked canal can result in two totally different behaviors of the climate and the vegetation, one healthy and the other unproductive. The water cycle is both the vocabulary and the articulation of the language of water.

A FORGOTTEN WORD

What it comes down to is that a single word has been forgotten when we settled on our current language of understanding and communicating water. That forgotten word is *transformation*.

The water cycle is the circulation of water through different states—of life, environment, and climate. The water cycle can be seen scientifically as the action of the physical laws of evaporation, transpiration, condensation, and precipitation that form the backdrop for planetary, biotic, and human activity.

If we give it the chance, then water transforms from liquid to solid (as ice or snow) and from liquid to gas (as from soil moisture through a plant to vapor in the air) in a temperature range perfectly accessible to our sun-warmed planet. We say that water is a language because we are constantly communicating with and influencing how these changes of state happen. We need to keep communicating with the landscape, encouraging water to enter into the soil and plants and trees to transpire the water into the atmosphere, for the clouds to form and the rainwater to be held in the land. If we don't tend to these actions that keep the water cycle functioning, then what we get is a desert on one hand or a flooded plain on the other, as nothing is there to hold the water when it falls.

Water is tremendously heavy in its liquid state (try carrying a full watering can to the garden!), yet this same water lifts effortlessly when transpired by plants into the atmosphere. Here, having transformed into huge clouds—myriad, cottony shapes, gently blowing animals and castles through the sky—it then transforms again into rain, releasing the load in a downpour of many millions of cubic meters of water to the ground again.

This water cycle—so fundamental to life—must be protected at its source. Its nature is to transform, keeping the world teeming with life. We need to understand transformation, transpiration, and transfer to make the worldwide changes to rebalance the climate. As we have isolated the process of burning fossil fuels, filling the atmosphere with too much carbon dioxide, these words in the language of water as an obvious remedy to our planetary climate crisis have been forgotten.

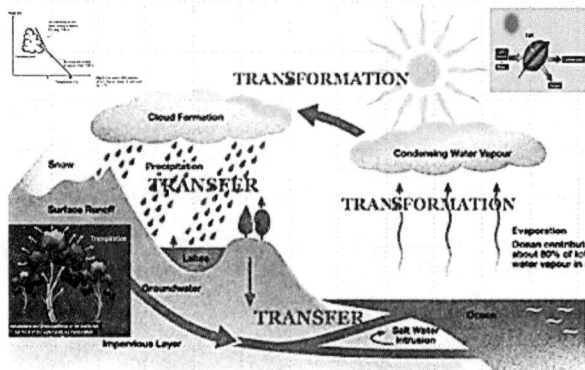

The water cycle. Courtesy of IB Environmental Science and Systems.

LOCAL ACTION'S IMPACT ON THE WATER CYCLE

On a River Ganges project we partnered on in India, satellite data was used to assess trends of temperature changes in conjunction with the water availability over a period of 20 years in the small project area encompassing 18 villages. We partnered with the Earth Sciences department of the Indian Institute of Technology, Kanpur, India, and the University of Leicester and Kings College in the United Kingdom to interpret and analyze the satellite data and find out whether the on-the-ground situation of the communities corroborated the picture that the data was giving. Part two of the project was to find out whether there were any questions the communities had that could be scientifically corroborated through the analysis of the satellite data and that could be helpful to the communities in terms of creating more bodies of water, reducing local temperature, and reviving a healthy local water cycle. The goal was to interpret the satellite data through the eyes of the community's experience.

What we found (the Bansathi–Bani effect) was a five- or six-degree Celsius difference in surface temperature in villages 30 kilometers apart. This difference resulted from the way one village worked on enhancing the water cycle, growing rice with water from a functioning canal system, and the way the other village deprecated the water cycle by not having access to the canal or creating the conditions for the water to flow.

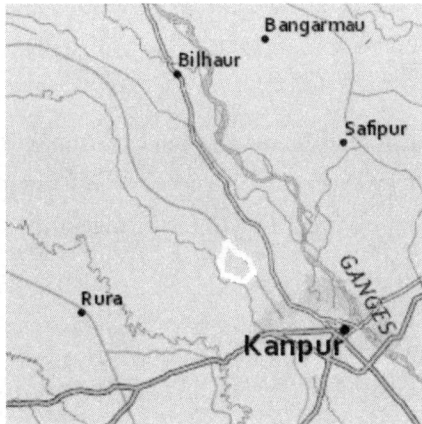

The topological base map of the region with the canal and the research area highlighted.

In the base map above, you can see the research area marked out in relation to the city of Kanpur and the River Ganges in India. There is also a prominent line top to bottom, left to right, parallel to the Ganges, indicating a canal system, running directly through the research area. The canal is not just a single channel of water but a network of main water courses supported by countless smaller waterways bringing the running water directly into the fields.

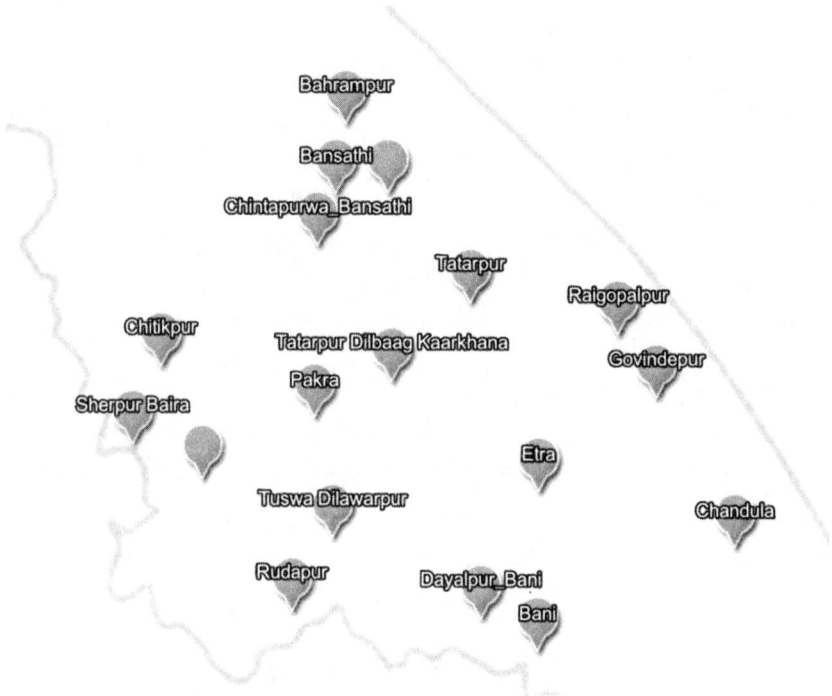

Eighteen villages of the research area.

To understand the different parameters impacting each other, we developed a system of layering key pieces of information onto a multi-dimensional map using the following information:

- the farmers' interviews
- the topological base map
- the crop patterns
- the temperature satellite data
- the vegetation health data

The data and the community experience matched perfectly. The farmers gave a qualitative analysis of the experience of temperature fluctuations locally that the data itself could not articulate. Final analysis from both the qualitative and quantitative data pointed clearly to the local action that helped the water cycle in one part of the region and demonstrated what action was needed in another area to bring the water cycle back to optimal functioning.

At first, when one tried to make sense of satellite data, it seemed complex and difficult to read. Then a surprising result came into focus. Where the canal was running and feeding through smaller channels into paddy fields for growing rice, the temperature was five to six degrees Celsius cooler than where the channel was blocked.

The local cooling effect of five to six degrees Celsius (an average temperature registered by the satellite) was brought about by providing running water to the fields and the evapotranspiration of water from the rice plants in the water-rich villages of the north as opposed to the wheat grown in the water-scarce villages in the south. (See the layer of satellite data showing the temperature difference in a band around the canal in the photo insert.)

As one sees from the series of maps we produced in October 2017 for the Kanpur area (located in the color insert), the path of the blue line in the background that corresponds to the canal also corresponds in the satellite data to a light green color, which translates to a temperature five or six degrees Celsius cooler than the red and orange areas around.

This difference was clearly visible when zooming in to the satellite temperature data within the research area between the north and south villages. (See photo insert.)

The satellite data map in the photo insert illustrating the vegetation health shows the same picture of huge variation in line with the canal.

We now returned to the locations of the farmers (shown by the different pins in the image in the photo insert) with whom we had carried out interviews at the beginning of the project. We wanted to present and discuss the difference in the temperature in the satellite data with the communities and its clear relation to the water cycle.

A TALE OF TWO VILLAGES

Bansathi (upstream village)

Community meeting at Bansathi.

In the northern communities in Bansathi, the farmers reported their good relationship to water. Water was plentiful there. There were many big ponds holding water, some with fish, and hundreds of bore wells indicating a water-rich underground aquifer.

The canal in this village is augmented with an abundant system of *bambas* (channels), *bambis* (smaller channels), and *nalis* (narrow waterways) that distribute the water throughout the vicinity of the canal. In the upstream areas, the canal and channels were being cleared and maintained periodically by the government, allowing the water to flow and be held continuously. The fields were also connected to the local canal system, allowing them to grow rice, a water-heavy crop.

In Bansathi, it was confirmed that all the water channels were working and facilitating groundwater recharge. Even wastewater from a hand pump was collected in an innovative way that allowed it to percolate back into the aquifer. The groundwater for the bore wells was accessible at 20–22 feet.

Bani (downstream village)

We then held a gathering with the farmers in the southern research area in the village of Bani.

Community meeting at Bani.

Here a totally different story emerged. There were three big ponds in and around that village. All of them were in a poor state. The ponds were filled with silt and had no water in them. In earlier times, the villagers used to dig out and clear the silt for construction and repairing their houses; without support from the government, however, that traditional practice had now lapsed. The channels were also blocked with tough growth and weeds, so no water flowed in them at all. The government had not come for maintenance work in years.

Blocked channel to canal from Bani.

The fields were no longer connected to the canal system as the overgrowth and siltation had claimed them long ago. The Bani farmers revealed a tale of long neglect from government officials and the villagers themselves. Monu Tiwari, the spokesman of that village, added, "Kyunki ye antim chor hai is ilake ka, yahan koi nahin aata, sab afsar upar se hi laut jate hain" ("Since this is the final stretch of the area, no official comes here and all the government officials go away having visited only the upper areas"). The village doesn't get any support from the Mahatma

Gandhi National Rural Employment Guarantee Act (MNREGA) or any other such schemes for creating water-retention structures.

This neglect had also led to depletion of the aquifer and decreasing water in the wells near the village. No farmer was able to grow rice in the village due to insufficient water. The groundwater level was at 28–30 feet.

Bansathi village had ample water through the efforts of the government and the villagers in keeping the water channels and ponds clear and full of water. Transformation of water to different states, the key action of a functioning water cycle, could happen effortlessly and continuously keep the village green, cool, and productive.

But Bani village had tried to access government money without success. Even though the villages higher up had been cleaned, the government workers had not gotten as far as this village, leaving the channels blocked and the ponds silted over. The data clearly showed that this lack of a working system of water distribution was having an adverse effect on the crop-growing potential. The silting of the ponds affected both evaporation and the recharging of the underground aquifer. The village folk here were facing multiple difficulties as a result.

In the community engagement meetings held in both these villages, the experience of the farmers was very clearly corroborated by the scientific data collected by the sensors and satellites. This was the first time farmers were able to see and hear how their experience was backed by hard science and shown to be relevant not just in their local area but also in the larger global context. The method of layering the farmers' experience with the topography and the satellite date gave a unique insight into the impact that running water had on their everyday lives.

Bansathi village is, at the moment, better prepared to deal with extreme weather events, by making sure—through caring for their water and keeping their bodies of water clean and silt free—that water has a prominent place in the priorities of the village. Meanwhile, in Bani village, where the water cycle has been neglected, there is an urgent need to restore their ponds and the channels, improving their local water cycle.

WEIGHING OUR ACTIONS

The use of satellite data overlaid on the farmers' practices gives a stark view of how the smallest actions are creating the climate situation of the planet. We are given the impression that our suffering from climate change has a single cause: burning fossil fuels. While the carbon dioxide excess in the atmosphere is of fundamental importance, of greater significance is that we as humanity have not come to terms with the degree to which we can upset the balance of the climate and the planet by neglecting to care for our water.

A community life of difficulty, lack, and uncomfortable surface temperatures can exist a few kilometers away from a parallel life of plenty, ease, and a cool surface temperature. That's the difference that caring for the water cycle can make.

REMEMBERING THE LANGUAGE OF WATER

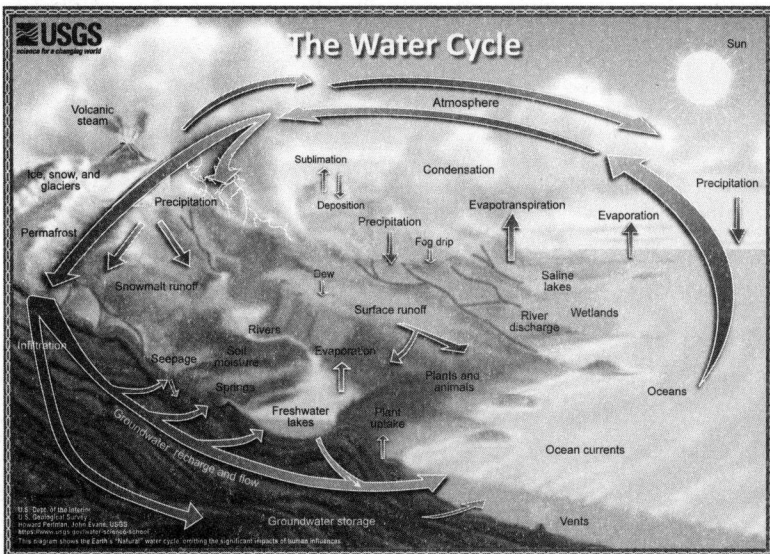

The water cycle. Creative Commons License public data from https://usgs.gov.

The water cycle is a continuous process of transformation, and in each of its four steps—evaporation, condensation, precipitation, and runoff—water moves and transforms. Thus the language of water is spoken on the planet.

Evaporation

For evaporation to happen, there must be surface water, such as in ponds, lakes, rivers, and snow. When water sources dry up, we lose a basic mechanism for water to move and transform in the water cycle.

Condensation

Once the vapor reaches the higher, cooler atmosphere with temperatures falling to zero, it condenses into ice and liquid droplets. The energy that was absorbed to evaporate the water is now released into the atmosphere, a perfect cooling system for the planet. Cities often become heat islands when there is no evaporation to cool the surface; the heat and dry air that rise up into the atmosphere burn off the water vapor in the clouds before it falls to the earth. Heat islands get created when we cover vast areas with concrete in cities or when a landscape is so devoid of surface water and groundwater that it has become a hot, dry desert with no moving water at all.

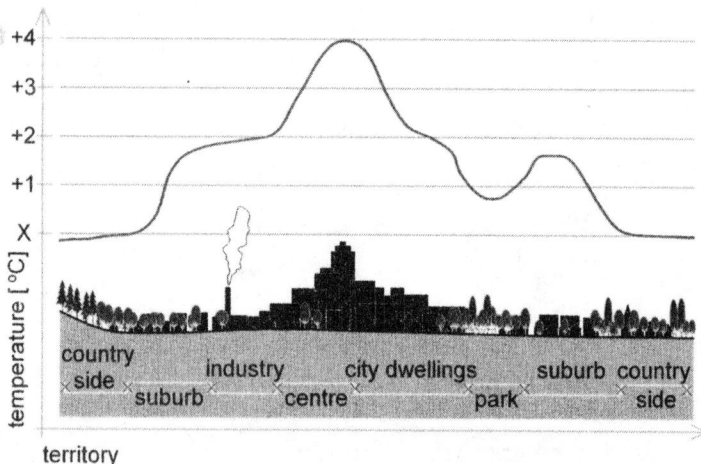

The increase of temperature familiar in the city on a summer day.

Why not simply turn this picture around? Planting trees and holding water in cities will turn the graph upside down!

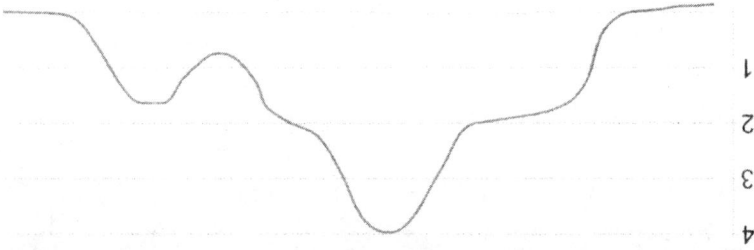

Cooling down.

Precipitation

What causes the clouds to transform into falling precipitation (so often the weather forecast incorrectly tells us whether that rain will fall or not) is still something of an enigma. But once the water falls from the sky, it needs to be held, slowed, and stopped so it can recharge underground aquifers and percolate into the soil without running away.

Runoff

Runoff is the water that drains away from an area and becomes lost from the water cycle on the land. Ideally, all the water falling on land should saturate the soil, then be stored in ponds, streams, or aquifers. Once the soil is saturated, water moves into a flow that eventually becomes streams, rivers, lakes, and oases. From here, evaporation begins again and the water cycle completes a turn, repeats, and the water balance is maintained.

The water balance indicates how much water stays locally within the cycle—between precipitation, soil moisture, vegetation evapotranspiration, and storage in the aquifer—and how much drains out of the water cycle.

This system of cyclical transformation of water is the foundation to all life on earth.

KEY COMPONENTS OF THE WATER CYCLE

All the key components of the water cycle have equal importance in allowing the language of water to be spoken. A primary component of the water cycle is soil. The soil begins by holding water so that different forms of life can transform the decay of previous seasons into the minerals for spring's new productions. The soil is perfectly textured both to allow water to penetrate into the ground and to hold water in its aerated spaces. The soil is like a sponge that concentrates life in a layer from which new plants and trees can grow.

Is the soil part of the water cycle or is the water cycle part of the soil?

The next key component is the trees. A tree begins by rooting out moisture from the soil. It then sucks up the water by internal pressure, up to 50 meters against gravity! This pure water is now able to react chemically to translate oxygen from the atmosphere into organic carbon through the action of photosynthesis in its leaves. The water also transforms into vapor by the leaves transpiring to cool the surface of the earth.

So is a tree part of the water cycle or is the water cycle part of the tree?

The most crucial of all components of the water cycle are the clouds. A cloud begins by holding together condensation, blowing across the sky as if it were light as a feather. The cloud is a moving aggregation of liquid droplets or frozen crystals, where the water vapor in the air is cooled to a state of liquid or ice. The cloud transports huge reservoirs of airborne water before falling as rain.

Thus is a cloud part of the water cycle, or is the water cycle part of the cloud?

Through this lens we can see more clearly into the water cycle. The soil, the tree, and the cloud are not separately existing elements that happen to be connected by the cycle of water between them. The soil, the tree, and the cloud exist only because of the water cycle. If one looks at Mars, where there is no water, neither will one find any of these features.

Such a thought should impel us to look at the water cycle anew. The planet does not possess a water cycle that has some vague connection to the climate through physics, chemistry, and biology. The planet has developed in the presence of water and is indistinguishable from the water cycle.

Is the planet part of the water cycle, or is the water cycle part of the planet?

This may be why, though we learn about the water cycle at school, it is hard to fix it in our minds, as we cannot separate the whole from the parts that form it.

IMPACTING THE WATER CYCLE

We think of the water cycle as some huge geographical phenomenon taking place beyond our own sphere of influence. But here, in the satellite data and the two practices relating to the water of the two villages described, we see how sensitive the water cycle is to our own actions. Even from a satellite in space we can see the outcome of the different attitudes to water in these two villages.

If everyone was relating productively to the water cycle, what would our planet look like from space?

Compared to the 1.5 degree Celsius global heating of the planet since preindustrial times, from 1850 to 1900 the water cycle provides a cooling presence of five to six degrees Celsius locally.

While we cannot do much about global warming as individual citizens, we can do plenty with our community about the water cycle's effect on our climate and landscape.

DESERTIFICATION AND REFORESTATION

By cutting down the forests, we stop the transformation of water from the soil into vapor. By drying out the soil, we stop the vapor that condenses into cloud and falls as rain from being held in the land. By drawing out vast amounts of water and depleting the aquifers without replenishing them, the wells dry out. Through these local actions, we stop the water cycle from functioning and doing its job of hydrating the planet.

If we want to remember how the nature of life on earth is not guaranteed but is only the result of the daily miracle of transformation in the water cycle, we need only head to the Sahara and look at where this work of transformation has stopped. Nothing is there except sand and the burning sun in a senseless heat, which at night, without mediating forms to hold it, is exchanged for a bitter cold. It is hard to imagine that this is what soils and trees become when water is absent from the landscape.

In reverse, there are some examples in future chapters where this cycle is turned around.

The Flow Project's current project ReForesting Baran[1] details how a huge area of 125 square kilometers can become forest once again by collecting the rainwater, improving the soil, and planting trees.

Yacouba Sawadogo of Burkina Faso received a Champions of the Earth Award for stopping the encroaching desert in his area by regenerating thousands of trees and bringing the local water cycle back.[2] The potential is there to do this at a huge scale—not for profit from water as a commodity, but to help transform the downward degradation of landscapes into upward spirals of renewal.

Each community has its own ways of holding water. In the Sahel, on the boundary of the Sahara, each drop of water must be saved for the stimulation of vegetation growth. The zai pits on the edge of the Sahara are small depressions that hold the rain and are fed with compost into which crops can be directly planted. The scant rain is held to sustain the plant. If the water cycle can begin with something as small as a hole in a marginal desert area, the planet is our canvas. By looking after the water cycle and enabling it to function, it takes very little time for trees to start growing, vegetation to return, and nature to transform before our very eyes.

TRANSFORMATION REMEMBERED

Water transforms life when in harmony and destroys when fragmented. We are the conductors who must manage water in the catchment to create a health that we can easily take for granted until we are missing it.

The remarkable conclusion from this is that, instead of compiling endless reports on how our intellect can measure and mediate climate change, the practice of supporting the water cycle to function at an optimum level is already set up to dramatically rebalance the climate equation.

Of course the example above is just a local effect. Making a small calculation, the amount of carbon sequestered in a living soil, with ample water, and growing vegetation has been estimated using the Carbon Toolkit (see www.farmcarbontoolkit.org.uk) at 8–20 tons of carbon per hectare (depending on the temperature). Multiply this by the amount of farmland in the world, and you arrive at an annual figure of carbon sequestration of 40 billion tons, which is almost exactly the total excess of carbon our cars, planes, and industry are producing each year.[3]

As an example, The Flow Partnership, along with the Apricot Center UK, are piloting a project with the Church of England to hold water and regenerate the soil of their glebe (church-owned) lands. The retention of carbon in restored soils is being tested to help the church meet its targets of achieving net zero carbon by 2030 (voted for by the General Synod in February 2020, see https://www.churchofengland .org/resources/net-zero-carbon-routemap).

If we are serious about balancing our carbon emissions and reaching net zero carbon, then the example of the Church of England's glebe land at Dartington could easily be followed worldwide to accomplish this target through healthy water and soil practices.

A WATER SYMPHONY

One does not need to be a scientist to appreciate and understand the water cycle. Composer Ernst Toch invites us to try to understand the water cycle as we would appreciate a great piece of music. Instead of trying to piece together cloud, forest, soil, river, plant, transpiration, and air vapor as something we can utilize for a functional benefit, he invites us to treat these open potentials as chords that we are tasked with composing into a great symphony:

> Do you know about the hydraulic [water] cycle? It is so perfectly closed in itself that no one can tell where it starts. Chemistry calls water H_2O. But this H_2O is given in a continuous chain of situations, each a link between others. The cloud, the rain-drop, drizzle

and cloud-burst; the snow-flake and the single snow-crystal, sleet, the hail-stone, and ice, fog and dew, the spring, the brook, the stream, the ocean, vapor and steam—which is the real H_2O? It is liquid, solid and gaseous; it is without color, light green and dark blue. Even the rainbow is H_2O under certain conditions, in a certain situation. Put the tone in place of the molecule and you have the multiplicity of its appearances. Put the harmony in place of the drop and you get the multiplicity of its situations. What else is a 5-3-1 [chord] here than H_2O there? Both symbols are serviceable tools for a certain approach.[4]

When a community sits down and tries to imagine an arid region transformed by water, they have to imagine what they think the future will look like. They become like composers ready to conduct the different resources into a water cycle symphony. You too are invited to become a co-composer of the water cycle through transformation via the different actions you take to hold water in your local landscapes.

5

United Kingdom:
A 5 Percent Future

The earth is increasingly experiencing extreme weather events. In the last few months of 2023 alone, flooding hit the headlines in Spain, Italy, Sudan, Ethiopia, Turkey, Peru, California, and Saudi Arabia. In Sydney, Australia, on February 9, 2023, 96 millimeters of rain fell in just one hour. The reasons for the frequency and ubiquity of these extreme events are threefold:

1. The global warming of the climate means the atmosphere can carry more moisture and produce more rain.

2. The rising temperature at the poles means that climate patterns have less variation, causing more prolonged periods of droughts and more persistent storms that carry heavier rain, causing floods.

3. The heat islands of cities and deforested lands tend to squeeze low pressure buildup between areas of high pressure, creating the conditions for these extreme weather events.

IS HIGHER RUNOFF THE GREATEST THREAT TO THE FUTURE?

Where does this huge amount of excess water falling from the sky go? It is estimated that, globally, 730 cubic kilometers of runoff water is drained and pumped away each year from the land into the sea.[1] Very soon after these floods, California, Saudi Arabia, and Australia will lack water and experience further droughts.

Why are landscapes across the world, forested just a few generations ago, now routinely becoming desertified and bare? Why do floods from storms possess a more ferocious energy, releasing new frontiers of devastation? Why is the climate increasingly so unstable and unpredictable? What has been the real impact of draining the wetlands (often done especially to make way for development and agriculture)? How does the depletion of aquifers and the covering of surfaces with concrete affect the balance of water, climate, and soil?

These questions speak of a general problem that is now visible all around the world. The challenge is not simply to reduce these extreme weather events, but to make sure the surplus water doesn't drain away so that we can hold some of it for the dry months ahead that are surely coming. By storing the water successfully, we can reduce future flooding and future drought to make a dent in reducing the incidence of these extreme weather events.

Understanding water as a flow that one can guide during a small window of opportunity when it falls as rain from the sky, before it concentrates into a disastrous flood event, helps us answer the questions that are challenging the planet at the moment.

When it starts raining, everything happens at once within the landscape. Rain sets off a whole cascade of processes that we often feel helpless about. When the water falls from the sky, we have literally a few hours to do something with it before it disappears out of our system. Or maybe we are further downstream and water suddenly appears. We can all do something with our rainwater runoff.

The community thrives when it has a stable, somewhat predictable, and secure relationship to nature. Our civilization is now experiencing anxiety and uncertainty because of global climate and environmental instability. This imbalance comes from our propensity to dominate nature, which of course has its own system and will not and cannot stay under our domination for too long.

The character of water has its own vocabulary of slow, store, filter, and sink that naturally teaches communities around the world how to keep their landscapes healthy and alive. The solutions explored here are simple and inexpensive. They stand as principles that can be applied anywhere. They do not require specialist knowledge, only an experience of how water behaves and a knowledge of the language of water in the landscape.

Learning to speak the language of water challenges and equips us to manage water in different situations and scales of landscapes. Anyone who has read of the success of Belford, United Kingdom, in stopping flooding using small and inexpensive local interventions in the flow pathways—often made with local materials like wood from the nearby trees—can be inspired to do something similar in their own water-challenged regions.

Location of the three catchments.

We compared three different-sized catchments in the United Kingdom, looking at what has been successfully done and what still could be done to restore the water balance there and understand some of that language of water and its behavior in the landscape.

BELFORD BURN, UNITED KINGDOM (SMALL CATCHMENT)

The village of Belford in Northumberland, with a population of about a thousand, had a long history of flooding from the nine-kilometers-long Belford Burn, disrupting life on at least five occasions in the four years before 2007. (*Burn* is the word for a small river in the northeast of England and Scotland.) The catchment of Belford is six square kilometers and ranges from upland pasture to lowland arable farmland.

The UK Environment Agency proposed constructing a concrete pond to hold the water on the edge of town before it entered the village.

Even this, with the magnitude of storms experienced, could not guarantee the village's protection. Additionally, such a feature would have cost the Environment Agency well over £2 million and was not justified by the small number of households it would help.

Since such high costs for building a traditional flood defense scheme could not be justified, the local Northumbria Regional Flood Defense Committee funded the implementation of a soft engineering scheme to construct dozens of flow intervention structures called bunds in the catchment upstream of the town. This was the Belford Catchment Solutions Project: a partnership between the UK Environment Agency, Newcastle University, and the local landowners and farmers from the village of Belford.

Twenty-six interventions that trap sediment, improve water quality, create new ecological zones, and slow the flow of water (each one an aspect of the language of water in the landscape) were trialed and built to successfully hold water upstream from Belford during the next flood event. The principles applied were to intercept, slow, store, filter, and sink the water.

The bunds, or interventions, created online ponds (on the course of the river) and offline ponds (adjacent to the river). These ponds work by storing water when the river is high and releasing it slowly back to the river after the peak has passed. Bunds were also built across the overland flow routes. These bunds intercept fast flow pathways of flood water, preventing runoff from reaching a water course too quickly. Large woody debris and other features were also installed to slow the flood peak and divert it onto the floodplain. Local residents corroborated that by 2014 flooding had "stopped completely in Belford."

By applying this strategy of collaborative water management, the outcomes included

- reduced flood risk downstream;
- reduced levels of pollution;
- habitat created for birds, wildlife, and aquatic creatures;
- increased biodiversity; and
- increased farm productivity.

The actual construction cost of the Belford project was between £70,000 and £100,000. This soft engineering solution, in harmony with nature's principles, addressed two major problems. It brought the local landowners, the council, the environment agency, local residents, and scientists together to arrive at a long-term, effective solution to flooding in that area. It also arrested the runoff of the best and most fertile soil of the area, improving the area's ecology and fertility.

Building interventions at strategic places in the landscape to retain water upstream is a catchment management strategy that is used very successfully by communities around the globe. It is an effective, small-scale, sustainable, and community-level approach for flood risk or drought-prone areas. Working in collaboration with nature, this soft engineering approach helps reduce flood/drought risk in riparian areas at a fraction of the cost, time, and effort of implementing only large, hard engineering solutions. Above all, it brings back the responsibility and engagement of local people in managing their local water effectively rather than relying only on government-controlled schemes, which often don't happen due to lack of adequate funds (as the cost of such schemes could run into several millions of pounds).

RIVER DART, UNITED KINGDOM (MEDIUM-SIZE CATCHMENT)

The River Dart is 75 kilometers long and its catchment drains 475 square kilometers of land. There were some dry months in the summer of 2022 and then again in February 2023, with prolonged periods of rain in between, bringing above average rainfall for the year as a whole. Yet in April 2023, well before the onset of summer when the reservoirs should have been full from that vast quantity of winter rain, South West Water, the company responsible for the catchment and for providing water to the residents of the area, put a ban on using hosepipes for watering gardens in the area served by the Dart reservoirs.

Our own experience of living alongside the river Dart for 20 years evidences the neglect of the whole catchment. We have walked the catchment and seen the changes. The Dart begins as a tiny trickle coming out of the ground from the marshy land on top of Dartmoor

National Park—a wild, desolate, and high catchment area of around 365 square kilometers in the southwest of England. As the trickle enlarges in breadth, the landscape funnels the water from the surrounding hills into the single flow of the river careening through the landscape 30 miles down into the town of Totnes and then eventually into the sea at the town of Dartmouth.

Five years ago, a £3.8 million engineering scheme addressed the threat of potential flooding to the town of Totnes by raising the wall around the river Dart by one meter right across the town.[2] This scheme aimed to protect 400 houses in the town from flooding. What it did not do, however, was address the whole river system—from upstream in the wilds of Dartmoor where the river originated to downstream where the river emptied into the sea. They only made a fragmented intervention where the river met the town of Totnes.

Just in 20 years of our living near the Dart, a change has become evident. The water nowadays tends to drain off too quickly from the compacted soils (stamped down by cows and hardened by machinery). The rainfall upstream (in Devon there is a lot) immediately causes the river waters to rise downstream, as all of that rainwater tries to drain away at once. Immediately, following some rain, there are local instances of pathways along the river becoming submerged in floodwater and becoming unusable. Once again, the real problem in the catchment is how to slow the water upstream so that it does not flood downstream, while at the same time storing enough water in the reservoirs for year-round use. This principle holds whether the situation is a drought in an Indian village, a flood in a UK town, a drought in the Sahel, or a flood in Peru.

The one meter of added height to the wall along the banks to stop the town of Totnes from flooding causes even more rainwater to drain out to sea during the life of a storm event. When a storm comes, the speed of the river rushes at 1–3 meters per second, with the heightened walls providing an extra draining capacity of approximately 2,000,000 cubic meters of rainwater out to sea over the course of 24 hours.

The capacity of the reservoirs around the Dart[3] is 115,388,000 cubic meters. In a dry year such as 1995, the level can drop below 40,000 cubic meters. Creating more storage capacity in the upper Dart catchment using a similar approach to Belford's would have addressed both the issue of flooding and the issue of a future drought at the same time without the need for raising the wall throughout Totnes. The rainwater would have been held upstream rather than simply draining through the river system and threatening the downstream town of Totnes with floods. This solution would have been preventative as opposed to mitigating and—as we saw in Belford—a far more cost-effective solution for the longer term.

This is what we mean by learning the language of water: taking action along all the different stages of a river and understanding the ripple effects of those choices. Doing it this way requires quite a different mode of attention as opposed to hard engineering, which, while needed, often focuses only on a single problem such as raising the height of a flood defense wall that has predictable results. When we understand the language of water, we can experiment to understand the behavior of water and the influence of features throughout the river system in the catchment. One becomes an artist, an engineer, and a scientist, using one's imagination to put together all the different areas of the landscape into a single workable, successful solution in which everyone living locally has contributed to a successfully resolved water situation.

This catchment-wide approach requires a different style of engaging. When farmers, local councillors, and the community talk with each other, collaboration emerges to discover the natural solution to the problem—combining farmable terrain with features slowing the water on the land instead of draining it. For a fraction of the £3.8 million, both flood and drought could be addressed by allowing better natural slowing and storage upstream, and the ability to hold more of that rainwater in the reservoirs as well. Now, in the summer of 2023, the whole area is looking dry again.

RIVER EDEN, UNITED KINGDOM (LARGE CATCHMENT)

The River Eden in Northern England is 145 kilometers long and its catchment area is 2,400 square kilometers. In December 2015, during Storm Desmond, Honiton Pass in Cumbria recorded 341.4 millimeters of rainfall in just 24 hours, which means that approximately 900,000,000 cubic meters of rain fell in the catchment area in those 24 hours.

What happened to that rain?

In the case of Storm Desmond, the water could not penetrate the already saturated ground and ended up pouring into the city of Carlisle downstream, flooding it and causing vast destruction.

The cost of cleanup alone after the 2015–16 floods was considerable, as is shown in the following table:

Impact category	Best estimate (£ million)	Low (£ million)	High (£ million)
Residential properties	£350	£308	£392
Businesses	£513	£410	£616
Temporary accommodation	£37	£31	£43
Vehicles, boats, caravans	£36	£31	£41
Local authorities (excluding roads)	£73	£55	£92
Emergency services	£3	£3	£3
Flood management asset and service	£71	£63	£78
Utilities – energy	£83	£75	£91
Utilities – water	£21	£16	£26
Transport – rail	£121	£103	£139
Transport – roads	£220	£165	£275
Agriculture	£7	£6	£8
Health	£43	£32	£54
Education	£4	£3	£5
Other (wildlife, heritage and tourism)	£19	£13	£25
Total	£1.6 billion	£1.3 billion	£1.9 billion

Cost of Storm Desmond and other storms in the United Kingdom 2015–16.[4]

The Environment Agency officials said the Cumbria flood defences did work, but, no matter how substantial any defenses are, "you can always get water levels higher than that, in which case it will go over the top."[5] While this is of course true, it also suggests that working with nature by slowing the flow is a safer strategy than trying to control the floodwaters after they have collected under our doors.

An idea mooted by the Environment Agency was to make Carlisle secure from a similar intensity storm in the future by creating 16,000,000 cubic milliliters of extra storage capacity outside Carlisle (approximately the size of four football stadiums). This "solution" was never carried out as it would have been wildly expensive and almost impossible to find appropriate locations for such huge structures. The decision then was to continue seeing and treating floods as a downstream problem, separate from the upstream destruction of the water cycle through deforestation, compacted soils, and excessive drainage.

Paul Quinn and Mark Wilkinson, pioneering hydrologists in the United Kingdom working with the James Hutton Institute in Scotland as well as the duo behind the Belford Project, along with The Flow Partnership, presented a proposition to DEFRA (Department for Environment Food and Rural Affairs) that the water storage should be held in multiple small, local structures upstream all along the catchment, which would limit the peak of the water rise and result in water being distributed across the catchment rather than becoming a devastating flood downstream. The idea was to hold the water upstream as the rain fell, before it gained momentum and became a torrent of floodwater. Our solution was to reroute the future disaster cleanup funds by preemptively building flood-mitigation measures upstream at a fraction of the cost.

However, DEFRA is still looking into embracing a catchment-wide approach that would restore the relationship between rainfall and holding water in the landscape. It seems harder to put in smaller amounts of necessary funds before a storm event and far easier to spring the massive cleanup costs after the floods have happened (not to mention the widespread misery caused by floods, with lives upended and lost, which can never be assigned a cost). The reasoning is always that there is no evidence that the small water-holding solutions will work in a large catchment. If you don't learn the language of water and work with it, you'll never find out, and if you don't try it, you won't know success or failure firsthand.

The first step in directing water upstream is to use the imagination to see how the water flow can be managed optimally according to a whole solution for the catchment.

After walking the catchment, we allowed for the free thinking needed to join up all the different locations where interventions could be made into a whole catchment design.

In any catchment, the designs are drawn up along with the community, imagining the landscape together, with necessary interventions, preparing to hold temporary water storage over the whole catchment during a storm event. In the case of a drought, designs are also drawn up to imagine water storage interventions over the whole catchment to recharge and replenish the underground aquifers of the entire region.

DESIGN INTO PRACTICE

The unique part of the story in Belford was that the village received money from the local council to turn these whole catchment designs into reality. This was an excellent opportunity to test how the designs worked through subsequent storms.[6] The map of the Belford catchment in the color section shows the area from the upstream hills (at leaft) to the downstream village (on the right).

It is helpful in the context of this chapter to reflect on the design, success, and possible improvements of the individual features, where each separate intervention reflects the strategy of the whole catchment. (Refer to the Belford catchment map in the photo insert.)

Next, we take a walk with Paul Quinn through the catchment of Belford to see the individual features, before considering their overall effect.

Wooden Barriers

A wooden barrier diverting water into the forest. Courtesy of Paul Quinn and his team.

This feature lets most of the water out from underneath, but when there is a really big storm, the water backs up and spreads into the forest. Paul Quinn illustrates the nature of experimentation in natural design, suggesting that with hindsight he could think of even better alternatives that might improve the original. Here are Quinn's assessments:

> This particular ditch with the wooden barrier is situated in the forest. These interventions could be better designed, with natural materials, and maybe made slightly higher. That was our first attempt. We all have to have practice. Remember practice, practice, practice. Don't be afraid of making mistakes. And then the right design comes along, resolving the problem we set out to tackle in the first place.

The wooden barrier holding back water in times of flood. Courtesy of Paul Quinn and his team.

Offline Whisky Barrel Ponds

An offline pond. Courtesy of Paul Quinn and his team.

The offline pond holds water for the duration of the storm, to be diverted to less saturated areas once peak flow has passed. It holds the water during peak flow and is designed to divert the water to where it is less flooded to slow the torrent of floodwater. According to Quinn, "this is the most characteristic feature in Belford. We found this beautiful location in a field, and I said, 'That is where we need to build a dam that is designed to leak.'"

Half the water that comes in has been diverted from the channel, and the other half flows in from across the land. By building this wooden bund to one meter at its highest point, with the wood extending one meter into the ground, at it stores about 600 cubic meters of water, collecting in from all angles of the land. The structure temporarily holds the water that leaks through around the edges and then the water makes its way slowly back to the channel.

Here is the same structure when it's dry, when it is still farmland:

The whisky barrel feature returned to farming. Courtesy of Paul Quinn and his team.

Quinn reports:

> The farmer was very keen that he should not lose any land since there was such marginal farming in the area. This structure is also engineered with double bracing, so that the cows can scratch themselves on the barrier but not knock it down. Meanwhile, it keeps a lot of sediment from flowing away. The farmer is happy. The cows are happy. There is a small impact on the farming business but a very big impact for the people of Belford since the flooding has stopped!

The Permeable Timber Barrier (Bund)

A wooden bund in a field. Courtesy of Paul Quinn and his team.

Many people ask why there is a random fence across the field. When it rains, two rivulets of overland water flow this way. Even though, by all appearances, the fence looks out of place on a sunny day, when it rains heavily, the soil is degraded and produces lots of runoff, which can be slowed by this bund.

The wooden bund holding and slowly releasing the surplus of rain. Courtesy of Paul Quinn and his team.

Here is Quinn's assessment: "The picture shows the structure shortly after a storm. It slowly fills up around the edges and then returns to the channel, holding the water only temporarily. The idea is that when it is full, it can hold the peak surplus of the storm and 12 hours later the pond is gone, and the field is back to being a farm field."

Woody Debris Leaky Dams

Further down the system, there is a little forest where we put a series of leaky dams with large woody debris.

Leaky dam with woody debris. Courtesy of Paul Quinn and his team.

Quinn reports that "at first, the local agency said we could not cut down any trees. But then they did allow us to cut down some sycamore trees growing nearby. We planted more indigenous species of oak, alders, and ash in their place. It looks very natural but it is an engineered structure. You can see the logs are all holding each other in position and they are attached to the bank. This backs up all the water onto the floodplain, which is needed to make the leaky dam work."

Soil Bund

Quinn's assessment:

> As we follow the river down the catchment towards the town, it is very fast flowing. So we built this structure which is made out of soil and materials from an old quarry.

A soil bund allowing water to collect. Courtesy of Paul Quinn and his team.

You can see a pipe in the structure. We always use a big pipe as the structure is designed not to fill up but to leak, so that it can work in very big storm events. That is a wheat field for that year.

This image shows what happens during the storm.

Surplus and water and sediment are kept in the field behind the bund. Courtesy of Paul Quinn and his team.

We built this dam as a road, almost like a bridge, so the farmer can traverse backwards and forwards. You can see from the colored water how a ton of sediment came out during this rain event. And the farmer attempted to plough this sediment back into the soil. This is where you get the bumper crops, for this is where all the nutrients are. So the structure ticks many boxes with multiple benefits.

Buffer Strips

A one-meter-high bund within a buffer strip. Courtesy of Paul Quinn and his team.

Quinn reports:

The farmer was in a government funded environmental scheme called a buffer strip scheme where they were asked not to build in the first 10 meters from the ditch around the back. This left abandoned land. So we built a bund and scraped material from in front of it to make the structure about a meter high. We usually do not go higher than a meter, as this keeps them within any required permissions and they are simpler to build. When it rains, water is taken out of the channel to fill this feature up. Again it contains a pipe to allow the water to drain through.

The bund holding back water. Courtesy of Paul Quinn and his team.

A series of nature-based structures, speaking the language of water in the landscape (intercept, slow, store, filter, and sink) not only stopped the flooding in the Belford catchment but also improved the quality of life for the farmers and their livestock upstream and the town's residents downstream.

In 2015, the enormous Storm Desmond affected all the villages in Northern England and the River Eden, flooding Carlisle. Belford and Pickering, two smaller catchments where natural interventions had taken place, did not flood. Of course this does not prove conclusively that the method works, but see if you find the following reports, given by Cronin's evaluation of the impact, convincing:

During the recent storms in Northern England the Belford scheme operated and protected property from flooding, with only one case of minor inundation. It is clear that the risk has not gone away as flooding has been close on a couple of other occasions. The research has shown that in Belford the natural flood management techniques have reduced the frequency of flooding and are very effective in small to medium-sized storms.[7]

Belford has not flooded since the construction of the natural water-holding features. This design was worth its weight in gold to the community and the work was finally recognized with an Institute of Civil Engineers (UK) North East Robert Stephenson award, given every year by the Institute to celebrate civil engineering excellence in the region.

Geoffrey Lean, an eminent UK journalist, wrote in the *Independent* describing a similar project:

> Stuck at the bottom of a steep gorge draining much of the North Yorkshire Moors, Pickering was flooded four times between 1999 and 2007, with the last disaster doing £7 million of damage. The solution, its people were officially told, would be to build a £20 million concrete wall through the center of town to keep the water in the river. No one thought it was ideal: it would have impaired Pickering's attraction for tourism. But then they were told that they could not have it anyway since too few people would be protected to satisfy the cost-benefit analysis for such schemes enforced on the Environment Agency by the Treasury.
>
> At that point—as Mike Potter, chairman of the Pickering and District Civic Society puts it—the townspeople were "spitting feathers" and decided to take matters into their own hands. Hearing from a local environmentalist how the moors had traditionally released rainwater much more slowly—and of how, centuries ago, monks at nearby Byland Abbey had built a bund to hold it back—they decided to try to go back to the future.
>
> They got together with top academics from Oxford, Newcastle, and Durham Universities to examine all options. Much the best plan turned out indeed to be to try to recreate past conditions by slowing the flow of water from the hills. Impressed by the intellectual endorsement, official bodies like the local councils, the Environ-

ment Agency, the Forestry Commission and even the Department of the Environment, Food, and Rural Affairs (DEFRA), joined in.

They built 167 leaky dams of logs and branches—which let normal flows through but restrict and slow down high ones—in the becks [the word for streams in English spoken in northern England] above the town; added 187 lesser obstructions, made of bales of heather and fulfilling the same purpose, in smaller drains and gullies; and planted 29 hectares of woodland. And, after much bureaucratic tangling, they built a bund, to store up to 120,000 cubic meters of floodwater, releasing it slowly through a culvert.

After 24 hours of rain, just three months after it was inaugurated, Mr Potter climbed up to the scheme and found it working well. Then he went home, "switched on the TV, and saw all the floodwaters elsewhere."

He adds: "While there was devastation all over northern England, our newly completed defenses worked a treat and our community got on with life as normal." The total cost, he says, was around £2m, a 10th of the original wall which, he believes, would not have coped with the Boxing Day conditions anyway.[8]

The focus on a catchment-wide holistic solution worked to manage the flow of the water. The peak of floodwater that used to rush down the river and flood the town is now spread out so that the rate of flow in the river is now manageable.

What will it take for the lessons of Belford and Pickering to be taken up and widely applied? One can prove what a wall will do but not say exactly how a leaky dam will behave. The language of water works by inclusively designing with nature to optimize the balance of water in the landscape. It does not work by separating the river from the town with a wall, as an isolated act that can be checked, monitored, and verified into a fact.

THE 5 PERCENT FUTURE

Sometimes critics of natural catchment management say that it is too difficult to negotiate with farmers. It is much easier to sign a contract with an engineering firm. But what is it we are asking the farmers to do?

Surely these examples from around the world show that farmers want to optimize the water in their landscape.

The combination of features works to hold the water in the field and the landscape where it can also solve the droughts that come a few months after. The principles are simple. Farmers put 5 percent of their land aside to slow the flow on each field. They also allow for a short period of inundation that quickly drains after 12 to 24 hours, still allowing a crop to grow or animals to graze.

The 'treatment train' approach
Examples of holding water measures and their placement

Sustainable Drainage Features: swales, bunds, ponds and grassy filters.

Buffer Strips: where designed to hold water.

The 'Ditch of the Future': the prime location for holding water and recovering lost top soil through erosion.

Small Headwater Floodplains: storing water, recreating wetlands, woodland, woody debris and new habitats.

The 5% Future
5% of land out of production
And

5% of floodplains for temporary flood storage

http://research.ncl.ac.uk/proactive/5future/

Just 5 percent of the land is needed to put the runoff attenuation features together into a whole catchment solution.

A 5 percent future of local land-use adaptation can give a 95 percent improvement of whole catchment resilience.

The 5 percent future design for land shows how this work can scale up from the size of Belford to restoring the landscapes of many larger catchment systems with simple features that work with the flow of water and nature.

Much of our thinking has the tendency to try to solve a problem once and for all with a singular solution. If we leave our thinking open to imagine how we can better optimize the whole catchment by optimizing the flow of water, we can use our imaginations to change the experience at the level of the individual farm, a small catchment like Belford, a middle-sized catchment like the Dart, and a bigger catchment like the River Eden, ultimately scaling up to the management of planetary catchments as a whole.

Belford/UK

Zishavane/Zimbabwe

www.theflowpartnership.org

Ditches of the future at Belford (left) and Zimbabwe (right). Courtesy of The Flow Partnership.

- **Principles**: Intercept, slow, store, filter, and sink; slow the water upstream to prevent flooding downstream.
- **Methods**: Wooden barriers, offline ponds, bunds, leaky dams, buffer strips, natural design and engineering.
- **Further applications**: Small tributaries, bigger catchments as the River Dart or large catchments as the River Eden.

When we first learned of the success in Belford, we were surprised to see that some of the interventions to stop flooding were the same as the interventions used in India to manage drought. Called johads there, the principle of slowing the flow of water so it can soak into the ground and replenish the aquifer is universal. Although the context determines how to apply the language of water, its principles are universal.

One of the arguments used by policymakers is that, though the interventions can work in small catchments, there is no evidence that they could work on big catchments. In India, by building thousands of medium-sized ponds (holding 10–40,000 cubic meters of water each), whole rivers in catchments the size of the River Dart have been revived.

A DESIGN TEMPLATE FOR SOME FEATURES IN THIS CHAPTER

Large Woody Debris	Target: Peak flow in small ditches and channels
Positioning	Large tree trunks spanning the width of the ditch, 30–40 centimeters above the typical pre-storm flow depth. Choose location where there is potential for out-of-bank storage.
Construction	Tree trunks are usually securely pinned in position on the bank and are oriented to cross over each other to aid stability. Roughen the riparian zone/floodplain with trees and spare brash. See Wilkinson (2010) for more details on woody debris management.[9]
Storage Capacity	Difficult to quantify owing to leaky design. Features attenuate rather than store and are often constructed in series.
Maintenance	Some small scour (removal of sediments) is allowed after construction resulting in settlement of the trunks. Some movement expected in flood conditions, but downstream movement must be prevented. Inspection is recommended after large events.
Cost per feature	Felling and positioning costs are £100–1,000 per feature.
Comments	Brash from felling is placed on the floodplain. Trees (holly and hazel) are planted on the floodplain to increase roughness.
Lifespan	Limited lifespan of a few years to 10 years. However, frequent inspection and vegetation management may be needed.
Design guidelines	It is important that the features are securely held in position and can withstand being overtopped. Permission to fell trees (sycamore in this case) must be granted, with the trees replaced with indigenous species (e.g., oak). Low-growing species are preferable to increase riparian zone roughness (e.g., holly and hazel).

Large woody debris. Quinn P., O'Donnell G., Nicholson A., Wilkinson M., Owen G., Jonczyk J, Barber N., Hardwick M. and Davies G. (2013) Potential Use of Runoff Attenuation Features in Small Rural Catchments for Flood Mitigation; NFM RAF Report.

Offline ponds	Target: Peak flow in small channels, plus local overland flow pathways
Positioning	Riparian zones, in a buffer strip or sited in the corner of fields
Construction	Draw off channel with an inlet armoured with coarse material to prevent scour, soil bund at front of feature with a suitable drainage pipe. Wood can also be used.
Storage Capacity	300–800 cubic meters. Some opportunistic sites may allow for more storage (200–4,000 cubic meters), but the volume should not exceed the 10,000-cubic-meter capacity limit covered by the Reservoir Act.
Maintenance	Check for scouring of inlet feature. The soil barrier may erode but should stabilize after grass has established. Sediment may accumulate to the level of the pipe and may need removal.
Cost per feature	£5K each but cost can be greatly reduced if several are constructed in close proximity.
Side effects	May be some local waterlogging and loss of crop production.
Comments	It is recommended that the maximum bund height should not exceed 1 meter and that grass cover is established as quickly as possible. There are additional habitat benefits associated with the feature.
Lifespan	Lifespan is estimated at 10 to 20 years, but inspection is needed. Grass/weeds should be cut at least once a year.

Offline ponds	Target: Peak flow in small channels, plus local overland flow pathways
Design guidelines	Choose an optimum location with the farmer. Permission for draw-off channel construction is needed. A skilled digger driver will compact the bund and thus it should not erode. Drainage pipe is usually quite large (roughly 30 centimeters for a 500-cubic-meter pond) to allow the feature to empty in 8–10 hours. Make sure to design a spillway; this can be engineered with rocks or may use the natural field landscape toward the edge of the bund.

Offline pond. Quinn P., O'Donnell G., Nicholson A., Wilkinson M., Owen G., Jonczyk J, Barber N., Hardwick M. and Davies G. (2013) Potential Use of Runoff Attenuation Features in Small Rural Catchments for Flood Mitigation; NFM RAF Report.

6

Africa Unfiltered

It is a six-hour drive south from Harare, Zimbabwe, to Mazvihwa in Masvingo province where the Muonde Trust Center is located. We attended the World Food Day celebration and the first ever three-day, in-person, all-partners Water School Africa and Flow Partnership meeting there in 2023.

The Water School Africa (WSA)[1] is a continuously developing partnership between communities engaged in natural water-retention practices in their local villages and regions across the African continent. It holds online learning sessions to stimulate community interaction and learning in sustainable, local, community-based water resource management, enabling communities to share their successful methods and techniques for holding water in the ground to become water secure.

The view from the window of the car is of shrubs growing on semiarid land seemingly devoid of farms, and it feels like we are looking out on a country whose complexity is completely hidden. Our Zambian partner who is driving us, Mugove Walter, grew up here and has known this area since his childhood. He gives us a background of its troubled history as we drive through.

When European colonists came to Zimbabwe, led by the commercial trader Cecil Rhodes in 1890 in the form of the British South Africa Company, they claimed all the land and cattle as a reward of conquest.[2] Over time, they replaced the traditional African homesteads with huge estates run by settler farmers that made Zimbabwe the breadbasket of Africa. Zimbabwe (or Rhodesia as it was known then) used to supply much of southern Africa with agricultural produce. When Zimbabwe became independent in 1980, these big estates were disbanded and the land was returned to the Indigenous Africans to resettle on their properties, in many places becoming small, self-sufficient farming endeavors once again.

Before the settler farmers took over, the indigenous practice was to farm the land within a tradition of respect for the nature of rock, water, soil, and life, manifested in a spiritual system of belief that joined the modern day to the ancestors. The colonial era dismissed those belief systems and overwrote those practices, chopping down many trees indiscriminately, including the sacred muonde trees, to clear the land for growing commercial crops and to use the wood.

The indigenous practice of cultivating wetlands and planting the crops sensitively according to the availability of water in any year was exchanged for the colonial system of draining the land, introducing the plough, and growing crops in planned rows on managed land.

In terms of farming and looking after the land, one could say there are two sides to the coin. On the one hand, the indigenous practice was small scale. Village communities farmed the land, producing food for themselves and their communities, with individual freedom as to how each family, group, or tribe interpreted its relation to their spiritual heritage and the nature of the land itself. There was no one set way of farming, and the impact on the earth was light.

The colonial system imposed a single idea of agricultural productivity on everyone—seemingly a global practice in all colonized places—and to enforce this with penalties and punishments. The water was drained from the land, which was turned into a resource from which wheat and vegetables could be grown and sold to nearby South Africa and across the continent, and the soil was artificially enriched with fertilizers.

As Anna Brazier, a sustainability consultant in Zimbabwe, notes in her piece on indigenous knowledge:

> In 1890, when the white settlers arrived in the land now called Zimbabwe, they saw that local communities had developed complex ways of living with a variable climate and environment. Farmers were using shifting cultivation. A diverse mixture of drought-resistant crops (including millet, sorghum, cowpeas, and groundnuts) were planted in fields cultivated with hoes. Land-use was governed by traditional leaders. Religious and cultural systems evolved to conserve natural resources.
>
> During colonial times, from around 1927 until independence, European farming techniques were introduced, including clearing

trees from fields, mechanical ploughing, and the planting of single crops (monocultures). Resilient, indigenous crops were thought to be inferior and discouraged, while maize (originally from Central America) was promoted. As time went by, farmers were encouraged to buy expensive inputs, including hybrid seed and livestock breeds, fertilizers, herbicides, and pesticides. Wetland agriculture was banned, even though it had been practiced sustainably for thousands of years.[3]

By the end of colonial rule in 1980, these methods of the settler farmers had left the land depleted, with large areas deforested, the soils impoverished, and the landscape dry. When new, government-funded schools were introduced in each community, Mugove Walter introduced a curriculum that promoted learning by observing and working with nature—he has been doing this continuously for over 40 years. Through setting up a national and regional organization called SCOPE/ReSCOPE,[4] Walter developed a curriculum for holding rainwater and growing now globally popular permaculture gardens, so classrooms could be outside and children could learn from the land. In this way, his methods combined elements of indigenous practices of learning from the land alongside colonial systems of schooling.

Zachary Makanya, another Water School Africa partner, was also traveling with us in the car. Zachary leads the Rural Initiatives Development Programmes (RIDEP)[5] in Kenya, communicating traditional practices that help local farmers cope with the changing realities of the climate. The job of reacquainting the local communities with their own indigenous practices after their colonial past is as vital in Kenya—which saw colonization during the 1880s during the scramble for colonies—as it is in Zimbabwe.

The night before this journey, in a rare moment of getting an internet connection, we had picked up on an alarming news item in the *Guardian* (UK) reporting on an increasing cholera epidemic in the district we were traveling to. Our imaginations ran wild; we prevaricated about the safety of going at all. The unusual, brooding weather with heavy, unseasonal rains lashing at us throughout the journey added to the tension, though we decided to go ahead on assurance from our local partners that the outbreak was quite far away from where we were going to be.

As we arrive on the outskirts of a town called Zvishavane, which was the last town before our destination, the gray of the sky is reflected in a particularly damaged landscape. In shock, we hear it is a huge asbestos mining town. We are driving along a rain-filled track that curves around close to a gigantic stone factory that looks as if it has come from the heart of an industrial city. Asbestos was once ubiquitous in the construction industry for its multiple strength-giving and fire-resistant properties. In the 1900s it was conclusively found to pose an acute health risk, causing mesothelioma in the lungs from breathing in the dust particles that were fine enough to enter and fatally damage cells in the body. All those working with asbestos were prone to completely avoidable early illness and death. Asbestos was banned in many countries and is gradually being removed, at great expense and by specialist teams wearing protective clothing, from all buildings where it had been used—or so we thought until we saw this mine. And yet here, in Zimbabwe, under the banner of money and jobs for the locals, the mine was alive and kicking and releasing large, invisible quantities of those same lung-destroying, fine asbestos dust particles.

On what basis is the Zvishavane asbestos mine functioning, still mining and selling asbestos even when it is a known killer for those who breathe its dust, and a poisoner of soil and water?[6] It is breaking the code of life—allowed to exist for profit only, and under the pretext of creating jobs yet poisoning everything living that surrounds it. As we bump along in the car around potholed tracks that have not even been maintained by the mine, the mine seems to offer no prospect of hope for the planet, its people, or its water.

There is a sense here that the colonial mentality has achieved an absolute hold over the planet and its resources by having no qualms at all about exploiting the land and the people as it maximizes what can be extracted from the earth for profit. To accept the deal on offer requires accepting this destruction and exploitation of the earth in exchange for the reward of earning a living—no matter whether your life is long or short as a result.

After a while, we seem to be lost and have to ask for directions to the Muonde Trust Center in Mazvihwa. Following some vague directions from a passerby, we retrace the route past the mine before heading along a track that we hope leads toward our destination. For a long while we feel a sense of doubt as this deserted pathway bumps along without any signposts, markers, or people to ask the way, the landscape holding only dry trees as far as the eye can see. We seem to have left normality behind, driving along in a limbo of unknowing, trusting we are on the right track.

About an hour or so later, just as we were debating whether to turn back, the Muonde Trust sign appears. We soon learn we have arrived at a remarkable place with quite a unique history.

ZEPHANIAH PHIRI

The person who planted the seed for setting up the Muonde Trust and Water School Africa was an Indigenous Zimbabwean called Zephaniah Phiri. Born in 1927, Phiri was always a rebel. As a young adult working on the railways, he fell afoul of the colonial authorities for political activities; after a period of being jailed, he was blacklisted from employment. Starting in 1966, without ready work or access to fertile land, Phiri could do nothing but subsistence farm on almost completely barren land to support himself and his family of six. Undaunted and with a tremendous spirit, he began observing the rainfall that fell on this barren piece of land and its patterns, experimenting with wells, ponds, infiltration pits, and other small-scale water-holding techniques to restore the health of his land. In colonial times, the white colonizers had imposed contour ditches to manage water during heavy rainy seasons. These contour ditches were not actually level, but were dug to slope downward, thus draining the water and the rich topsoil off the land. As a result, the contour ditches both caused flooding where the runoff was directed and dried out the land from which the valuable water had been drained, silting the rivers, which continues to be one of the reasons why there are few flowing rivers in Zimbabwe today.

Zephaniah Phiri went against this practice of draining the land by creating pits (now famous as Phiri pits) where the water could enter the ground from these ditches. Phiri also argued that contour ditches should be level with the land, rather than sloping downward. Sometimes, where rains are excessively heavy, ditches can be angled to drain the excess water, but the area of South Zimbabwe that Phiri lived in was exceedingly dry. The colonial formula that had to be followed under threat of punishment was completely out of place. Thanks to Phiri, the ditches are now pegged so that they have zero slope and are called dead-level contours. Instead of draining away, the water in the contour ditches sinks into the ground and replenishes and recharges the water table to ensure moisture for cultivation.

To make sure these contours are dug on the level, an A-frame called a *chidhonge* (meaning "little donkey") has been popularized by the Muonde Trust. The A-frame, used since the times of the ancient Egyptians, carries a loose hanging weight between two poles that allows the contour to be mapped out with pegs. Amusingly, the colonial application of the A-frame was not to assist the farmer in finding the level of the contour, but simply to measure that the ditch was the requisite two meters wide.

A-frame for pegging contour lines for ditches.

In the beginning, it was tough going for Zephaniah Phiri, with the colonial system of sloping ditches still reigning supreme in the minds of people and the fear of reprisals if the system were changed. He was arrested three times for "farming a waterway" (planting trees and vegetables in the water ditches) until a magistrate eventually demanded to see his land. The magistrate was so impressed that Phiri's ponds and earthworks were reducing rather than increasing runoff and erosion that he ruled against the government's land development officer and granted Phiri resource rights to use his methods of conservation farming in his wetlands. In 1973, a more progressive land development officer brought local farmers to see Phiri's drought-beating methods and his work became more well known. But even then, the safety his fame provided was short-lived. Phiri's beliefs, going against the proclaimed colonial practices, soon landed him in trouble again. During the Liberation War in 1976, Phiri was arrested, put under house arrest by the authorities, severely tortured, and even held for long periods in leg-irons until the end of war.

Phiri seemed to have entered into the secret (or not so secret!) language of water, understanding the transformation it brought to the land and to him and his people. His rallying cry until the end was "All this water that falls within my land should be kept into the soil. I call it planting water!"—a true mantra for all the water work in communities across the entire planet. Phiri taught farmers to know water and how to flow with it according to their own lands, without caging that knowledge within a fixed system.

But why was Phiri, whose water-holding techniques had so infinitely proved their value on the land, so disrespected by the colonial powers, who inflicted repeated punishments and restrictions on him? In the 1870s, various explorers starting with David Livingstone (1813–1873) and Henry Stanley (1841–1904) had shown that the interior of Africa was crisscrossed by rivers that could be used as navigation routes to the ports for the trade of goods. Until then, no foreigner had ventured into the interior of Africa. Upon a calculation of extractable wealth enforceable by guns, the next 30 years would see the whole continent of Africa carved up between the five powers of England, France, Germany, the Netherlands, and Portugal with not the slightest permission from the native populations or a discussion of how the colonial model was supposed to work for them. Unlike other colonial endeavors, where the land and practices were gradually assimilated into a new status quo, the colonial plan in Africa was worked out and deployed from offices in distant cities of the world with scant regard for anything else other than the control of resources and profit.[7]

Zephaniah Phiri was one of the few people who loudly voiced the contradiction between the colonial and indigenous systems of land management. It must have come as a surprise to some of the colonial bureaucrats that Africa was not to be farmed as if its land were a sunnier form of England. The administrators could not countenance that an Indigenous voice was pointing out the discrepancy in their methods. Their response was to always try and muzzle him and his spirit—unsuccessfully, we may add.

Zephaniah Phiri. Courtesy of Muonde Trust.

In the run-up to independence in 1980, Phiri redeveloped approaches to wetland farming at his home in Msipane, a neighboring region to Mazvihwa. Although he found sympathy, and his well- and dam-making skills were much sought after, no farmers were prepared to engage in water harvesting in their fields. They remained fearful of the authorities in any "interfering" with water. Only after independence did a newfound courage drive farmers to adopt the methods. In 1986, Phiri started the team that eventually became the Muonde Trust and spent several years encouraging similar innovations in Mazvihwa until his death in 2015.

Zephaniah Phiri stuck with the struggle and stayed true to an internal belief in the connection to land and water, even when colonial authorities were intent on censoring and punishing such sentiments. One can find resonance between Phiri's life and the story told in *The Life and Times of Michael K* by Nobel laureate J. M. Coetzee. Coetzee describes the life of an African named Michael K who survives by growing pumpkins in abandoned areas of land during the apartheid era of the 1970s and 1980s. When he is chased off even those areas by the white South African rulers of the period, Michael K ends starving with a single seed remaining in his pocket: the source of all of his hope and dignity. The book shows the contrast between the system that unfairly punishes Michael K and the bounty of nature that can regenerate abundance from a single pumpkin seed. At the end of the book, all that was necessary to live was found in the spoonful of water lifted from the well: "He would lower [the spoon] down the shaft deep into the earth, and when he brought it up there would be water in the bowl of the spoon; and in that way, he would say, one can live."[8]

TWO WAYS OF FARMING

The indigenous African perspective was to work with the water in the ground, retaining it, promoting wetlands, and learning techniques to naturally manage the rainwater when it fell. This was overwritten by the colonial settlers when they came. Their perspective was to drain the fields of water first and then return any water needed for farming in a controlled manner through irrigation. The colonial approach allowed much more extensive agricultural production, but in time it exhausted the soil, depleted the water table, caused droughts, opened the way for the natural water cycles to break down, and increased the risk of flooding in extreme weather events.

Is there something between these two methods of managing water in the landscape that expresses a more fundamental understanding that can be learned for future practice?

The indigenous practice was based on holding water on the land. The people encouraged wetlands and chose their crops according to the water available at the start of the growing period. If the rainy season had provided ample water, they would grow more rice; if only scant water was available, they would grow maize. The African farmer naturally understands water first, from there the soil, then the trees and crops that the land and soil can afford to support. It's not an understanding learned from books but an indigenous intelligence that comes to them from living on the land the year-round. Nature constantly speaks to them, teaching them what promotes harmony and what does not.

The colonial system of agriculture saw water as simply one variable among many others that fed into calculating land productivity. European settlers introduced the ox plough so that the soil could be prepared in neat rows over much larger areas, replacing the hoe, a traditional African hand tool. Because the hoe was used by women and the ox plough was worked by men, this also changed the nature of farming. As the emphasis moved toward productivity, the soil was enriched with artificial and chemical fertilizers. Water was drawn from boreholes going down 70 feet to irrigate monocrop production at an optimum speed that increased the quantity of produce and ease of harvest.

PHIRI'S SUCCESSOR: THE MUONDE TRUST

The successor to Zephaniah Phiri's work is the Muonde Trust, which was officially set up in 2014 by his long-term collaborator and an original member of Phiri's team, Abraham Ndhlovu. Abraham Ndhlovu's face is the epitome of one that has learned much through experience. His questioning look weighs up the world without judgment. When he first met Zephaniah Phiri, Abraham was one of those skeptics of Phiri's claims about the importance of holding water on the land. It was only when visitors started coming to learn from Phiri that he realized there was something fundamentally important in his practice and teachings and decided to spend time and learn from him.

At a pivotal time, 1986–87, some years after Zimbabwe's independence, when rural life was trying to find new direction, Ken Wilson,[9] an academic from Oxford, came to research the area of Mazvihwa. Ken Wilson also married a local lass and for a time lived in the nearby village. Together with Ian Scoones of London and Zimbabwe Universities, they started a research project in which Abraham Ndlovu traveled on his motorbike from farmer to farmer to document their practices. This research provided a platform for Zephaniah Phiri's work to reach farmers and laid the groundwork for the foundation of the Muonde Trust. Ian Scoones and Ken Wilson developed lifelong connections to the area, and Wilson became one of the founders of the Muonde Trust in 2014.

The work documented in Mazvihwa assessed the state of the land by visiting all the farmers in the area and learning of their challenges firsthand, initiating an analysis of the contract between colonial and indigenous agricultural systems, and paving the way for new ecological centers to work with their communities using this research as a blueprint.

By speaking to the farmers and documenting the challenges they faced, gradually a picture emerged of the contrast between indigenous practices and the colonial system that helped them understand the confusion that took place after independence in 1980, when the large estates of settler farmers were returned to local smallholding farmers overnight.[10] Ken Wilson, summed up the foundations of the Muonde Trust in 2015:

> In common with many other semi-arid regions, the ancestral agricultural system in Mazvihwa was not based on extensive dryland farming but instead focused around the intensive use of natural wetlands—makuvi or dambos in the hills and majeke riverbank gardens in the plains. Elaborate systems of trenches and furrows were dug by hand with heavy diagonal hoes to manage water in these areas and create a matrix of wetter and drier areas in which rice and maize could alternatively do well. The approach enhanced resilience because in wet years the rice would do well and over-run the field, and vice versa with the maize in droughts. During the colonial period a concern with preserving water supplies and with constraining African commercial productivity led to the suppression of wetland farming . . . encouraging an extensive farming approach, which of course is difficult to sustain because of the breakdown in soil and water management.[11]

Scoones and Wilson wrote this overview of the state of Mazvihwa in a research paper:

> There is an urgent need for the upgrading and expansion of water supply in Mazvihwa and other nearby community areas. Most people are reliant on unprotected water sources that may be quite some distance from homes. These are generally old vlei

wells in the hilly areas and wells in the river bed sand (*mifuku*) on the plains. Similarly stock and gardening enterprises are reliant on scarce water sources resulting in long treks for water by cattle in the dry season and an immense amount of labor invested in vegetable production. Government and donor water programs have had some impact in the area since independence, mostly in extending the network of deep boreholes and wells. It is not surprising that people put improving water supplies at the top of their development priorities.[12]

And finally, in 2015 Ken Wilson summarized the forestry traditions that had been turned upside down:

During the first half of the colonial period, Mazvihwa was heavily logged in order to generate electricity and construction timber for the asbestos and gold mines to its immediate north. This still has lasting impact on woodland structure in many areas. Subsequently people experienced many years of colonial extension efforts, including arrest and forced labor, to have them reduce trees in and around their agricultural lands, including de-stumping coppicing species and removing shade and fruit trees in fields that are also associated with ancestral spirits and ceremonies.[13]

The colonial practices were imposed under penalty of imprisonment, and the methods were abstractly conceived without taking into account the nature of the land, which suffered progressive degradation from then on. There is a huge misunderstanding of colonialism as a progression to a more advanced scientific outlook that made the traditional practices redundant. It was, in many ways, the opposite.

The challenge with holding water is that water is both beneficial and dangerous if allowed to accumulate in rapid flows and flooding. The colonial approach was thus to dry out the land, protecting against the wild aggregation of water, and then to reintroduce irrigation from reservoirs or bore wells as a controlled supply. The challenge in traditional farming based on holding the water in the land was how to tame the flow in high rainfall events for a managed flow in low rainfall periods.

Scoones and Wilson emphasize this background in their research. "In all encounters, people have impressed on the research team the necessity for action. In all areas there were recurrent themes. Predictably in an area that suffers recurrent drought and is under increasing resource pressure these were the problems of water supply, the need to secure stable arable production (e.g., through the use of wetlands), the problems of drought shortage and grazing and the problems of deforestation."[14]

When it came to the research conducted in the region, Zephaniah Phiri emphasized the importance of targeted action:

> [T]he research [conducted in the 1980s] wanted people to give their own problems in their daily life, so that people could indeed have time of bringing together their needs. This then brought the people of Mazvihwa to share their ideas together. They were able to say they need cattle, for their cattle had died during the drought. Farming land had also been a problem. They had no water for cattle and even water for drinking or watering gardens; they had no green in most part of the area due to water problems.
>
> So with all these findings people of Mazvihwa have been able to meet and say: "we need a well, a dam, a sandy well and also gardens." It was then I also answered I could help if there were any people who would need my help. I was asked if I could sink a well . . . I said "yes!" So at these meetings people started calls for wells, dams and sand dams . . . That was my problem to start the work.[15]

After working informally for many years, in 2012 Abraham Ndlovhu, Zaphaniah Phiri, Ken Wilson, and others set up the Muonde Trust. Abraham took the name from the Muonde tree, which mirrors the qualities of Phiri's vision. The Trust writes:

> The word Muonde in Karanga/Shona refers to indigenous fig trees of this semi-arid region, especially the big, beautiful free-standing *Ficus sur*. This is a tree associated with fruits and birds and life. It is noted to particularly grow and thrive in places where there is water beneath the ground. It becomes huge and ancient and enduring. It is a tree that accommodates ancestral spirits as they

move between the worlds. There are prohibitions on cutting or harming the muonde tree, and they are not cut down when fields are cleared for agriculture (people resisted government efforts to make them do this for many decades). Its shade is one of the most preferred for meetings, ceremonies, churches, and for resting between bouts of weeding in the hot summer sun. The muonde are honored but they are also fun. Children love to play in such figs, climbing to get the fruits, and (in some species), gathering the sap for birdlime and locally-made chewing gum. Overall, then, the muonde is thus a connecting point for community, ecology, and spirituality.[16]

The Muonde Trust further elaborates:

Everything at Muonde is focused on transforming the experience of development from one driven through top-down, externally-derived resources and ideas (in which locals are a "target" and exhorted to "participate") to instead one that people themselves own and that encourages the bottom-up generation and sharing of practical knowledge alongside providing empowering training when needed.[17]

Muonde tree *Ficus sur*.
Courtesy of P. E. Bingham.

ARRIVING INTO THE PRESENT

We arrive at the Muonde Trust Center in the midst of the celebrations of the World Food Day, held annually by the UN Food and Agriculture Organization (FAO) on October 16. This year's theme is "Water Is Life. Water Is Food. Leave No One Behind." The Water School Africa has co-organized the event with a regional networking initiative (PELUM—Participatory Ecological Land Use Management, Zimbabwe) and the Muonde Trust. We enter the large community hut, where a hundred or more farmers are seated and the program of discussions and presentations is unfolding. As we enter the room, where a rhythmic chant and dance of the Rujeko Choir celebrates the work of the Muonde Trust, we are suddenly in a quite different Africa! The rhythm no longer reflects a relentless outer greed of taking indiscriminately, but an inner harmony and celebration of the innate order of the world. Whereas the mine was all externality of acquisition, the dance is all inwardness, a recognition of a shared humanity.

In this atmosphere, we introduce the Water School Africa, the spirit of which could be seen expressed by all of these farmers recognizing and voicing their traditional methods of holding water in the ground. The language of water is being sung as a rhythm everyone in the room knows and moves to. The World Food Day is understood here as a celebration of the harmony of farmers who produce healthy, nutritious, and plentiful food from their land, beginning with holding the scant rainwater that falls in these parts. In the afternoon, a group of farmers and partners visit one of the farmers who had trained with the Muonde Trust.

Learning Crucible: A Visit to Local Farmer Handsome Fundu's Farm

The Muonde Trust center has developed practices that allow local farmers such as Handsome Fundu to combine a deep understanding of the landscape with a wide familiarity with different water-holding features for the varying amounts of rainfall he can collect, as well as come up with his own innovations around water-holding structures relevant to his land.

What is noteworthy about the farms we visit is that there is no feeling that the farmers have been told by Muonde to hold the water in one particular way. Each farmer understands the landscape of his own farm and how to use the pits and contours to collect water for crops throughout the year in a slightly different way. Ken Wilson assesses the diversity and success of such practices:

> While the farmers of the hills focused on infiltration and management of lateral flow under the soil surface and its storage behind ponds in places where heavy clay bands could trap and hold it for later use, water harvesters on the clay soils focused on capturing surface flow. By 2011, some 7 percent of farm households on clayveld and 13 percent on sandveld had built these individual ponds, most holding typically 100–300,000 liters of water, and a revolution had begun.[18]

The farmers feel a sense of pride and ownership now that this knowledge has been reawakened for the benefit of their families and for the other farmers who learn and train with them. Farmers like Handsome Fundu help groups experience the language of water as it is uniquely spoken in this one precious grouping of ponds, spillways, ditches, and additional flow routes for the water to run off in heavy storms. Handsome shows us the interconnection of features that hold the flow and make his farm water-rich even in a very dry period.

The holding of the excess water from the rainy season that would otherwise simply run off the land and be lost is now held in the soil and groundwater table for access throughout the dry season. This allows the crops enough water to grow plentifully through the dry season. The work of Zephaniah Phiri and the Muonde Trust has been to learn from all these examples and develop different ways of holding water according to the context: contour ditches that Zephaniah Phiri adapted from the sloping colonial versions, larger ponds that could include fish, small pits for local recharge, sand dams that hold the water without evaporation from the high temperatures, earthen bunds that stop the flow of the water, and the careful design of the land to mix the growing of trees and crops.

This work reestablishes the relationship between holding water and the health of the landscape through active experimentation. The implication, not fully articulated, is that what is being relearned is also the relation between Indigenous African people to the working of their own land. In each of the farms we visit with Muonde folk, it is the farmer's family who are the focus as much as the systems they have devised to hold the rain in their land. The land is returned to health through inclusion of the human beings who tend to it. The farmers and their families relearn their own system of being on the land, not by intellectually critiquing the colonial system, but by regaining an understanding and learning from the land itself.

Ken Wilson describes the change that the Muonde Trust Center has brought to its farmers:

> Water harvesting in Mazvihwa was now being driven by significant economic opportunities to produce early season vegetables, as well as to enhance drought resilience in the rainy season of the principal crops and gardening (or farming) in the dry season with micro-irrigation. One of the households in Muonde's long-term sample (which used to monitor trends), namely that of Handsome Madyakuseni, had developed a two-pond system filled by local stream flow and (astonishingly) had built a five roomed house during the years of economic crisis from the sale of tomatoes and other garden crops that this supported.[19]

After the visit, we eat together, a simple meal cooked outside on the fire. The produce that we have seen during the day—maize, peanuts, and beans—is now the content of the meal that we share, rounding out the lessons of the language of water in the food that we eat.

The next morning, we see the world reflected in the child walking to school, the simple breakfast food, the colorful landscape, each a special sight uniquely presenting the possibilities in the day ahead. The day is set aside for the meeting of the Water School partners, the first time we have met in person, after two years of planning, presenting, and enabling the groundwork through online meetings. We talk creatively about the work each partner has accomplished under the umbrella of Water School Africa. We appreciate our host, Muonde Trust, as one of its pillars, exemplifying the potential and opportunity of holding water

everywhere it falls on the ground, and skillfully regaining the living wisdom native to the area by spreading that knowledge far and wide.

Abraham appreciates that the Water School Africa does not limit
what the water teaches to a simple lesson in meteorology or food production or managing climate change out of fear. The community is
proudly reworking with their own tradition of being with the land,
which expresses itself in all manner of ways—commitment to their rural
life and all its challenges and joys, participating in the collective flow of
life in the village, a sense of life finding its direction anew in the culture
of the land and harmony of the planet.

On the afternoon of the second day, we visit another of the farmers
who have trained with the Muonde Trust, Kelfas Hove. Kelfas Hove
is a more mainstream farmer, who has continued to use chemicals on
his land. He is proud to show us his ploughed contours, which bend
around the field to help the water sink into the soil. He also shows us
the government technique called *pfumvudza*, which looks like a variation
of zai pits (see chapter 9 on Burkina Faso), that holds the water that falls
in the neighborhood of each plant, a technique that has been accepted
by many of the farmers with mixed success.[20]

There are many voices that come together in speaking the language
of water, and each is worthy of being listened to. It would be an easy
mistake to think of the Muonde Center as simply using the language of
water in prescribed ways that can be standardized with proper supervi-
sion. The spread of this knowledge and these practices are instead like
the wild fig tree from which Muonde Trust takes its name, with roots
freely seeking the water to find out something more about life in relation to the landscape. The Muonde tree finds groundwater close to the
surface, and the site of its growing is taken as a good place to put down
a well. Similarly, the Muonde Center is a sign of the general health of
the area and the multiple systems of farming that can hold water in the
land in the ways that benefit them the most.

INTENTIONAL RECHARGE

The next day starts slowly, filled at first with contradictions tangling
up together indigenous and colonial mindsets in ways that cannot be

resolved. The indigenous ways strive to learn from life, while the colonial mindset leads only to the self-interest of one group or system over another. The indigenous heart and mind seek to enter into the fold of the world, to see how the whole can flourish.

It is in this spirit that, during the partner discussions in the late morning, Tsuamba Borgou, from Groundswell International in Burkina Faso and a WSA partner, brings us to the concept of "intentional recharge." "Intentional recharge" means we do not intervene in the ecological system for a short-term goal of enriched crops or water for animals, but for the long-term health of the whole catchment, changing its fundamental cycle from drought to vitality. How do we move from individual farmers doing their own work to speaking the language of water and applying that to the catchment as a whole? How do we emphasize collective participation, to which many diverse communities can reference their water-holding activities? The beauty of such a realization between us is that it has not come from an analysis of the parts, but as a recognition of where the work can naturally step from where it is now. Everyone feels comfortable refocusing the emphasis to holding water and intentionally recharging the water table underground in this way.

Our visit in the afternoon is to a third farmer, Pedzisai Maburuse and his family, in a remote, rocky area. This farmer has worked out the exact nature of the flood risk posed by increasing cyclone and storm events and has devised a pathway for the water to flow over the rocks, where it can be slowed and held in the land. As he shows us around, what started as a defence against floods around the family dwellings turns into the most abundant exhibition of fruits and crops, as if we were visiting some exotic garden in a temperate land. Farmer Pedzisai Maburuse is even experimenting with growing apples on his farm, which has everyone in fits of laughter alternated with wonder. The relation to the land is creative and suggests its own solution to the challenges of climate, location, and soil. One is continually surprised by the ingenuity of the interventions and the selection of trees and plants that have succeeded in taming this uneasy water situation between the extremes of droughts and floods.

A REVITALIZED IDENTITY

Now we have reached our last night. Sitting in a circle, dimly lit by fading solar lanterns, we eat a quiet meal together and discuss what we all can do in our own regions to bring abundance back to the planet. The sky is dark enough to let us see the Milky Way. The positions of the constellations are of course different from Europe. We sit in quiet wonderment and appreciation of the village and community life, the beautiful Zimbabwean silence and the rural night sky. In discussions the next morning, we ponder how the language of water we have learned is not just about tackling droughts and floods, but regaining a sense of confidence to freely rebuild our relationship with the land.

The community cohesion of the area tended to by the Muonde Trust has grown not just in terms of the tomatoes we see being harvested and collected for sale, but in the community finding an identity and learning how to live well in the world. We reflect with Abraham and Emmanuel, a councillor and member of the Muonde Trust who represents a younger generation. The world now needs community cohesion around water to tackle climate change more than it needs diamonds (or asbestos!) from the mine. This time around maybe it will be the Indigenous African farmers' turn to provide the leadership for this through the movement of sharing, learning, and "planting water."

Zimbabwe: The Spirit of the Land Reawakens

Looking deeply into rearticulating their relation to the land and their old traditions gives an impression of how the language of water is finding a voice again within African communities. After independence, the reallocation of the large colonial estates necessitated that new bodies of knowledge be made accessible for smaller farmers occupying the land as reference for their actions on the land. Permaculture and related practices provided a good starting framework for this. Permaculture provided a set of beliefs that brought together a variety of approaches of working with the land systematically and mindfully.

In the late 1980s, a permaculture enthusiast, John Wilson, saw that a lot of people were questioning the approach to western "development" and were coming up with alternative approaches to managing grazing areas, social forestry, and involving local community. His efforts led to the setting up of Fambidzanai in January 1988, an umbrella organization to connect and train practitioners working on the land in permaculture-friendly ways. Zephaniah Phiri, with his track record of managing his farm and his water system over many years using local indigenous knowledge and knowing the importance of water to Zimbabwe, trained many others through Fambidzanai.

To see how the centers are progressing today, we met with John Wilson in October 2023 and traveled with him to visit Rob Sacco and the Nyahode Union Learning Centre (NULC),[1] Julious Piti and the Participatory Organic Research Extension & Training Trust (PORET),[2] and Nelson Mudzwinga at the Sashe farms and Zimbabwe Smallholder Organic Farmers Forum (ZIMSOFF).[3] Among other challenges, these centers focused on three major concerns in particular:

1. How to keep water on the land while training the flow to be able to cope with extreme storm events;

2. How to recover indigenous plants and trees so they could play their vital role in supporting the holding of water on the land;

3. How to bring the water work into line with spiritual and wisdom traditions of the community.

NYAHODE UNION LEARNING CENTRE (NULC)

Rob Sacco, a man with a big spirit and a big vision, is the inspiration and founder of the NULC. Rob understood that holding the water in the ground meant that one would need to create a flow pathway that could deal with storm water during frequent extreme climate events such as the cyclones and storms that visited his valley. One had to put every centimeter of land in service of slowing the water flow before it aggregated in a damaging flood, destroying buildings and homes.

Over a period of a few years, they constructed earth core dams, weirs, swales, spillways, hillside dams, interconnected water systems, and soakaways; managed woodlands, orchards, and gardens; and built an extensive reticulation system consisting of pipings and storage tanks on the four-hectare site of NULC. This system has survived a number of cyclones, including Cyclone Eline in 2000 and Cyclone Idai in 2019. NULC has an agroecology/watershed management team of 11 people.[4]

Today there is a flourishing secondary school teaching 700 students of all ages at NULC. Meanwhile, the Nyahode Union Community Technical College carries out intensive training in carpentry, tailoring, building, mechanics, and agroecology. Rob Sacco showed us around the whole 42 hectares of NULC, describing in detail each water-retention feature they had constructed there. It is easy when seeing such a lush, thriving environment to forget how much imagination it took to ensure that rains were securely held and that the soil held moisture for the trees, grasses, and crops to thrive. Each intervention is carefully linked to the others so they can work optimally in tandem. Here are some unique techniques and methods we saw in this series of ponds and other interventions:

- There were leveled-out sports fields for the schoolchildren at the top of the land that could also become floodplains in high rain events.
- High up on the land was an impressive pond with a 10,000 cubic-meter capacity, hand-dug by the schoolchildren.
- The overflow from this dam was fed into a neighboring dam with a 2,000-cubic-meter capacity further down the hillside. This second dam was made excessively deep so that the highest flow imaginable could not overtop it. The water from this pond infiltrated into the water table. Sacco noted that, "If you want, you can semi-seal the pond floor by growing a thick crop of wheat at the bottom of the pond, and then once grown, you shred it down so it sits at the bottom of the pond. But we wanted the pond to empty quickly so did not grow the wheat."[5]
- In building these higher dams, five further dams, and five smaller ponds down the slope, Rob Sacco's imagination entered the force of the water to understand how to slow it down, resisting the force of the storm by creating safe places for runoff and enriching the land at the same time.

Once they understood how this worked, Rob and the team at NULC applied these principles in many catchments in their neighboring communities across the Nyahode Valley. The free flow of the water collecting upstream is slowed by obstacles in the path of the streams and, at the same time, piped to cisterns from which rubber pipes provide a ready supply to households and farmland.

Rob tells us about surveying a neighboring community-owned area for the renovation of its stream. When he explores upstream, he finds the obstruction of a gully with damage caused by Cyclone Idai in 2019. If nothing is done about it, the water will simply collect at one place with an uncontrollable runoff, risking floods and erosion of good soil. Rob imagines how to counteract this obstacle by the nature of water itself, building a solid weir and pond behind to make sure that the rain from a storm event can be contained within a reasonable flow. Another solid weir could be built to serve as a footpath across the stream for school-children attending the local primary school. Further leaky weirs can be constructed to divert excess water to irrigate fields. A seasonal spring pro-tected by a ditch to help collect the water could become perennial and connected through a pipe down into tanks accessible to the homesteads.

You cannot observe and address the damage of Cyclone Idai, the fields drying out, and the relation of the community to the stream as individual problems. Rob is not simply thinking about water as a com-modity. He thinks about water in connection to the cyclone, the school-children, the fields, the homesteads, and water's own qualities. In Rob's way of working with the land, the crops and the community are one, both receiving life from the stream.

When water is not controlled through a modern system, you must dynamically hold the water through its relation to the land. "Planting water," in Zephaniah Phiri's words, becomes as important in its own way as planting crops. The water is held in the land in such a way that it allows the crops to flourish year-round while also holding excess water in rainy periods.

Rob Sacco describes the situation from the experience of holding water at Nyahode:

> Zimbabwe has quite a short, sharp rainy season and then a long dry season, with the moisture slowly coming out of the soil until the roots are no longer drawing moisture and then the country browns off. When the water is held in the ground to recharge the water table, even in the dry times, as in October, the water never quite dries beyond the roots. And then by the time rain comes, it will recharge again. So the roots perpetually have water. Grasses remain green throughout the year. An area that used to brown off is now green throughout the year.[6]

We learn from these visits to appreciate the contract between the land, the climate, and the water, where even in the most difficult and apparently impossible environmental situations there can be harmony—the alignment of ponds, crops, ditches, people, and trees in which all the elements are held by each other. We leave such places totally optimistic that community projects, if practiced on a large scale, could also find a solution to the global crises we are in at the moment.

PARTICIPATORY ORGANIC RESEARCH AND EXTENTION TRAINING TRUST (PORET)

Julious Piti is one of the six initial volunteers who founded the CELUCT (Chikukwa Ecological Land Use Community Trust) Center in the next valley in Chimanimani. In 1996, he settled on a piece of land in a drier part of Zimbabwe with as little as 300 to 400 millimeters of rain a year, which sometimes fell in just two storms. Julious, along with the local community, painstakingly dug 34 dams that helped capture and store the water from these storms. He estimated a thousand cubic meters of water could be harvested for the dry season if they did this.

Examples of dams dug by PORET. The sides of the dams are planted with pawpaw, which can then access the water and strengthen the walls. Photo courtesy of PORET.

In 2006, Julious founded an organization called Participatory Organic Research and Extention Training Trust (PORET) to research agroecology in this semiarid zone. This community-based organization has designed and developed an extensive interconnected water-harvesting system of swales and ponds/small dams at their community learning center.

We visited Julious at PORET on a typical hot, cloudless day. We could see high mountains where the clouds drop all their rain before they reach the plains. The ponds turn the landscape into a cupped hand facing the mountain to receive all the runoff and to sink the water into the ground. Further off, the land is fenced so that grazing cattle and wild animals do not eat the grasses and young tree shoots, whose roots keep the soil porous and able to absorb the rainfall.

Many farmers in the area are increasingly taking up similar water-harvesting practices in a farmer-to-farmer learning process, facilitated by PORET.[7] PORET also encourages farmers to understand how grazing patterns can affect water retention in the landscape so as to limit indiscriminate grazing. They encourage controlled grazing areas for the animals, which has helped regenerate grasses at their center, compared to the bare hillside just outside the controlled grazing areas.

PORET works with 24 villages under the patronage of the local chief, mapping their areas and developing solutions in collaboration with the communities to hold the water. Julious also convenes the Chaseyama Permaculture Club, a group that had just 31 members in 2015 but had grown to 450 households by 2020.

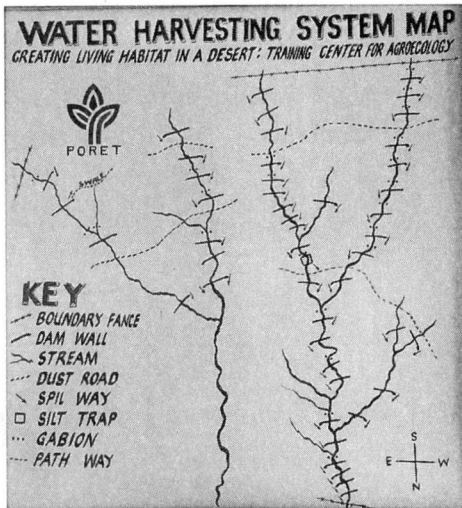

PORET research study for a catchment showing the streams and interventions of gabions, silt traps, and dams.

PORET selects indigenous varieties of trees for their beneficial relation to water. Trees hold water in the landscape and are pivotal in helping rainfall enter through the root system into the soil during the rainy season and drawing water up from the groundwater in the dry season. Planting the right trees for the land aids this passage of water to and from the aquifer. This contrasted with the colonial preference for gum or eucalyptus trees, high consumers of water chosen for their rapid growth and saleable produce even in water-scarce areas.

Julious shows us nurseries of macadamia nut trees, wild figs, and water berries that he grows to distribute to the farmers he trains and helps. These trees are chosen specifically because of their ability to tap into the water table to draw the optimum amount of water up through their roots.

Just as you can look up and see the relationship of the ponds to their source of water in the mountain, it is just as important to look down to the roots of the trees and see how they help bring water to and from the underground aquifer. The water from the rain on the mountain used to simply run off the land, and the deforestation of native trees left the aquifer's water untapped. The trees draw the water up from the aquifer and then evaporate it into clouds in the atmosphere—a prerequisite for a functioning local water cycle.

In each center we visited, the farmers have experimented with indigenous varieties of trees that help to draw water from the water table to the surface. The feeling we had standing before the nursery is that these trees are right for the land. They have grown up together with the land as one in the transformation that can manifest life. They carry traditional medicine, nuts, and fruits to eat, and are part of the land's lore. Julious recognizes these original species as allies in returning his land to health.

SHASHE FARMS

Another farmer we visited doing profound work with water is Nelson Mudzingwa of Zimbabwe Smallholder Organic Farmers' Forum (ZIMSOFF) and Shashe Agroecology School in the Runde Catchment in Zimbabwe. The Shashe farms were created as resettlement projects in 2002 on totally dry and depleted land.

Mudzingwa takes us around his 184 hectares of land with great pride. His neighbor and colleague Abumelech shows us his own family and farm on the same visit. Abumelech's farm highlighted that nothing leaves his land. This was clearly illustrated in the residue of every action:

the leftover plants after harvesting being put back in the trenches, the leftover residue after extracting oil from the seeds being used as green manure, and the rain held in ditches to become moisture for the soil.

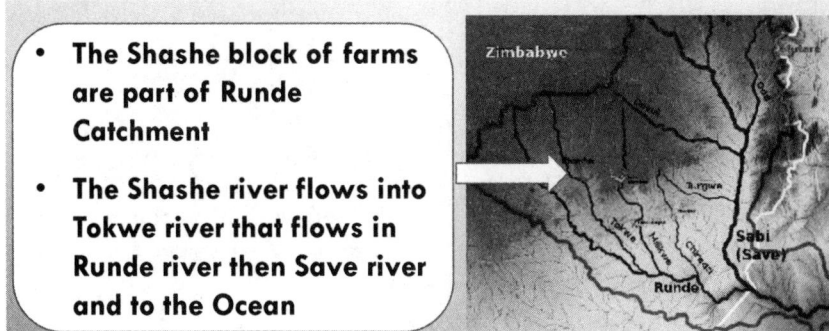

- **The Shashe block of farms are part of Runde Catchment**

- **The Shashe river flows into Tokwe river that flows in Runde river then Save river and to the Ocean**

The location of the Shashe farms. Courtesy of Nelson Mudzwinga.

In a talk given to the Water School Africa, Mudzingwa[8] related his experiences of restoring the water cycle through traditional methods using earth dams, dead level contours, stone barriers/walls, and interlocked keyhole beds as part of a social, ecological, and agricultural transformation. Nelson said:

We were also very much inspired by the work of the late Zephaniah Phiri, learning from some of his videos about Mr Phiri's best water practices and conservation . . . Coming into a new land, after resettlement into the Shashe area, was a big move. New farmers were coming, trees were being cut down, the water cycle was suffering with diminishing rainfall, and there was land degradation with a lot of gullies forming in different areas. Many farmers were wondering what the reasons were for this. When we came here in 2000, the Shashe rivers were becoming silted. We started discussing different ways of managing the land in lots of informal discussions, leading to proper awareness in each and every farm. We had lot of different trainings and set up demonstration structures to showcase living examples of helping the flow of water.

Everything in Nelson's story of the water cycle, from the collapse to the regeneration with the community, is realized through the simple observation of nature and the experience of doing the work. It highlights what can happen when the collaboration of local communities and their knowledge, wisdom, and successes are shared widely. The key slide in his talk was the following diagram of approaches to the work:

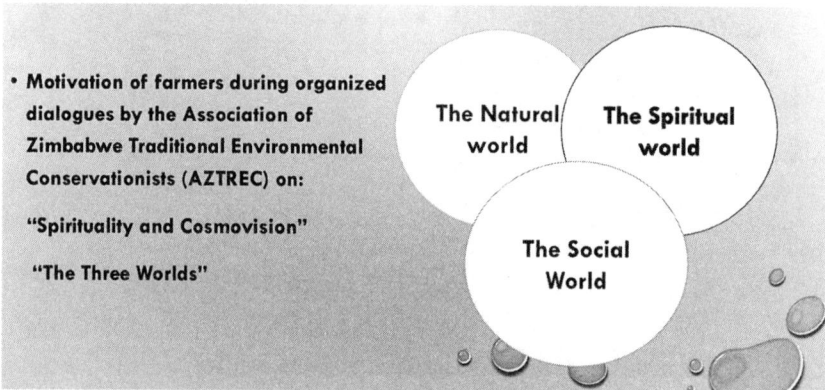

The collaboration of Nelson Mudzwinga with the Association of Zimbabwe Traditional Environmental Conservations (AZTREC). Courtesy of Nelson Mudzwinga.

In this threefold motivation for working the land, something is being clarified: The work connects the spiritual self with traditional ways to understand and care for the landscape. Working on this land, developing a relationship to water, is paralleled in how the relation to life is founded on an active relation to the spirit. The spirit in Africa is not something ethereal reached exclusively through special ceremonies and external pathways. The sacredness of the spirit is contained within daily life itself. Only by understanding this expression of the spirit inside do the external spiritual stories make sense.

Traditional wisdom is not a book that can be picked up and easily understood. It is practiced by relating to land, spirit, and community. This traditional wisdom does not contradict other ways of understanding. It is first a code of action that shows, through its practice, an alignment with the landscape and other beings.

An article by Takesure Mahohoma from Masvingo, Zimbabwe, details the following about the Shona people[9]:

> From a traditional point of view, Shona people knew and believed in ancestors who carry their petitions; ancestors that lie in the deep and unknown recesses of the past. . . .
>
> Apart from the belief in ancestors, the Shona also believe in a Supreme Being who is, however, unknowable. . . . This being is the final authority above and beyond the ancestors. At the top of the spiritual hierarchy was the god *Mwari*, described in eco-centric names like *Zame* (the unreachable horizon), and anthropomorphic names such as *Mariga* (all-powerful), *Chikara* (an awesome one), *Mutangakugara* (the pre-existent one), *Musikavanhu* (Creator of human beings). These varied names carry anthropocentric attributes of *Mwari* and reflect how *Mwari* relates to people. . . .
>
> Among the most visible aspects of Shona religion are sacred places and moments. These are fundamental elements of Shona spirituality. They are at the center of Shona people's view of ancestors, God, and the world.
>
> There are trees, forests, and insects that are sacred. They fall under providential nature and nourishment. Just like mountains and rivers, it is a taboo to cut down or destroy a sacred tree or forest. Sacred forests are traditionally called *rambatemwa*. Literally, these are sacred groves protected by the traditional leadership from domestic consumption and utilisation. The Shona believe that ancestral spirits rest in these groves. . . .
>
> Of all the key sacred objects, water is perhaps the most ubiquitous. Almost everywhere, the Shona have sacred pools, rivers, springs, and waterfalls. Evidence exists that the Shona believe that wells, springs, rivers, and other water bodies have been in existence from time immemorial. Among the sacred water bodies are wetlands (*madekete*), springs (*zvitubu*), and sacred pools (*madziva anoyera*). These water bodies are associated with autochthonous ancestral spirits.[10]

These individuals are not dedicating themselves so committedly to restoring these lands just to prove a point about farming—it is more about an inner completion in finding their place within their Zimbabwean home and landscape.

When Nelson arrived, the land was just grassland for cattle, if the grass grew at all in such a dry and degraded area. Yet in a single tree, Nelson was able to picture the transformation to come: in the place where he was setting up his farm, he felt affinity for a particular tree and felt moved to ask it for permission. Through the years he has been holding water on the farm, his relationship to that tree has grown and developed. The now majestic tree oversees a lush soil holding a variety of crops supported by ample water held from the rains. Nelson sees the health of the land represented by the well-being of that tree. His land becomes a story intertwined with the tree, its storyteller. Connecting with the farm is an open project that expands the horizons of the spirit, making articulate an ancient tradition.

The Shashe farmers—Nelson, Abumelech, and others—have built a community together on the basis of their innovations and traditional practices. The more they open up to the land, the more the spirit speaks to them and shows itself through their work. Nelson stresses the patience needed at the beginning to allow holding water to take root in visible forms before extending the idea into the communities. Once the farm was functioning and the water being held, only then did he let the work speak to other farmers, as the basis of their training. Maholoma writes:

> The life of the Shona has a clear distinction between the sacred and secular. From their spirituality, there are places and times when people experience the sacred. These places and times are celebrated at different levels. The experience of the sacred is a moment of close relationship between the living, ancestors and the unknown. At other times, it is possible to conclude that Shona experiences of sacred places are spiritually helpful because they are based on a mutual relationship with ancestors. The horizontal relationship with one another nourishes the vertical dimension of the relevance of the ancestors. There are taboos and beliefs used to maintain silence at sacred places. The Shona experience of the

sacred safeguards their morality, politics, and economic development. Therefore, Shona spirituality is underpinned by respect for nature because the latter literally and figuratively nurtures both the living and the dead.[11]

Indigenous cultures are often portrayed as relating to nature in a particular, fixed way. Yet when the Indigenous culture was challenged by colonial systems being superimposed, an entirely new way of being with the land was able to reassert itself. Each individual farm we visited was developing some aspect of the challenge of the individual farmer to reclaim a wholesome relation to his land. These pioneers have laid the path on which others can follow.

The spirit of this work and how it can be communicated far and wide is now in the hands of new generations of leaders like Daniel Ndlovu and Emmanuel Prince of the Muonde Trust. The cycle of research, literacy, action, and community integration is ongoing. The more the farmers are on the same page in their efforts, the greater the learning and the bigger the impact they can make together.

We have to choose—whether to stand outside a material system joined up by causal law or participate integrally in a living world. In the increasing extremes of climate that we are facing, the work of holding water is becoming even more valuable. These centers and their farmers are not passively following prescribed knowledge to decide their future. They are responding to a changing situation in relation to the water and the spirit of their lands. They are addressing a perennial question: "How can we look forward to a future of well-being for all living beings?"

8

Chimanimani, Zimbabwe: Community Guardians of Water

We are used to measuring climate change in increments, such as the increasing length of a dry season, breaking a temperature record, or the amount of rain that falls in one extreme storm event. At this point in time, we still have the capacity to respond to some of these climatic events that we see all over the news. However, where extreme circumstances concentrate to destabilize the whole ecological foundation of life, normality breaks down and the region and planet is torn from its normal course.

We were told about one such catastrophic climate event while visiting the Chimanimani district on the Zimbabwean border with Mozambique. In March 2019, Cyclone Idai tore through the region, causing extensive flooding over several days with almost limitless torrents of water. Landslides destroyed entire mountainsides, ripping up all the trees and vegetation. Hundreds of houses were lost along with thousands of human and animal lives. This kind of climatic event takes climate instability to an entirely new level.

Through extreme periods of drought, increased amounts of water in the atmosphere due to global heating, and sudden concentrations of low-pressure storms, a changing pattern becomes evident. Climate factors recombine in ways that can appear suddenly, overpowering the entire ecological foundation of an area and causing untold destruction as wrought by Cyclone Idai.

In this chapter we look at how the communities in Chikukwa villages of Chimanimani district, Zimbabwe, having understood the fearsome and ferociously destructive face of the language of water, have responded in the years after Cyclone Idai in 2019. The seven villages in the valley are repairing their ecological defenses, should another such cyclone occur in the future. Traditional ways to manage the livestock on the land have been plucked out of the past and are being attempted again. Newly imagined roles as "guardians of the catchment" to slow the flow of water and hold it upstream are being woven together into a community response to manage future cyclone-fueled destruction if it was to recur.

Cyclone Idai exemplified the changing pattern of severe climate events. It came to Chimanimani at the end of an exceptionally long period of drought, when typically the land fully dries out and massive heat is reflected back into the atmosphere. Sunlight is absorbed by plants through photosynthesis to transform carbon dioxide and water into vegetation, which actively transpires that water from roots into vapor and cools the earth. Cutting trees, overgrazing that produces bare earth, and erosion of soil all contribute to this process breaking down.

In the midst of such conditions, an initial tropical depression formed over Mozambique on March 5, 2019. This depression moved to Malawi and brought heavy rains and flooding with it. The progress and momentum of the tropical depression were determined by the extreme climatic circumstances prevailing in Malawi. With the unseasonal extreme heat all around, the low pressure could not dissipate but instead intensified its internal dynamic to a new level of devastating force. The tropical depression returned to the Mozambique Channel (the stretch of water between Mozambique and the island of Madagascar) and took on added force, transforming it into Tropical Cyclone Idai. An increasingly large air mass began to rotate around a strong center of low pressure carrying inward-spiraling winds and devastating rains. Tropical Cyclone Idai made landfall on March 14, 2019, at Beira on the Mozambique coast, bringing winds of 167 kilometers per hour and storm surges that pushed sea levels up 2.5 meters. It then moved westward, crossing into the Chimanimani district, producing an estimated 20 to 40 centimeters of rain over six days.[1]

In the immediate aftermath of the cyclone, on March 18, 2019, the UN Office for the Coordination of Humanitarian Affairs reported on the situation in Chimanimani:

The hardest-hit district Chimanimani remains inaccessible as heavy rains have damaged the roads and main access bridges have been washed away. At least 82 deaths and over 200 injured have been reported, mainly in Chimanimani, and 217 people are reportedly missing. At least 923 homes have been destroyed. In Chimanimani alone, eight bridges have been destroyed. In Buhera, the Marowanyati dam has overflowed and many families are displaced. Crops and livestock have been destroyed. Overall, more than 2500 households (12,500 to 15,000 people) are estimated to be affected.[2]

A Medicins Sans Frontieres vehicle is stopped by the floods. Frieden, M. (2019) The First 6 Days after Cyclone Idai in Zimbabwe; Voices from The Field March 29th 2019 "Medecins Sans Frontieres" https://www.msf.org/cyclone-idai-zimbabwe-first-six-days

In the district of Chimanimani, an additional 115,000 people were affected by flooding when the Nyahode and Charleshood rivers burst their banks. The rough dirt- and rock-filled road that winds down through the mountains to the Chikukwa villages (the only road in), difficult at the best of times, became impassable and had to be closed off (and remained closed for many weeks).

The Chimanimani Mountains mark the eastern boundary of Zimbabwe, stretching for some 50 kilometers and forming the border with Mozambique. Several rivers have their sources in these mountains. According to the 2012 census, 134,939 people live there, mostly rurally, in 3,500 square kilometers of land. The abundant rainfall supports fertile soils, natural forests, and diverse vegetation cover. The primary challenge is soil erosion and flooding, exacerbated by the mines that have been opened to access mineral deposits in the area.

From the town of Chimanimani, the spectacular (if bumpy) road drops from a good height to the valley floor and reascends into the mountains. After the one-and-a-half-hour ride on this unpaved road we arrived to visit the CELUCT Center (Chikukwa Ecological Land Use Community Trust), which has been at the forefront of the community response to the damage from Cyclone Idai since 2019.

Even after five years, the impact of the cyclone is still visible everywhere. The numerous landslides still show their unhealed brown scars of bare earth across the mountains. We could see hundreds of the new roofs provided by the United Nations, painted green (instead of the usual brown), identifying where the original coverings of the houses had been blown away by the cyclone.

However, we were told that the most lasting and important effect of the cyclone was that it brought the community together to work out how to defend itself against such concentrated forces in the future. As a community, they have completely rethought what working with the ecology, landscape, and water in their area entails and what speedy, adaptive, or preventative action they themselves can take should another cyclone come.

CELUCT is a community-based organization founded in 1991. Even before Cyclone Idai, high rainfall in this region often led to erosion, landslides, droughts, and flooding. The organization began looking into whether holding rain in the landscape could provide a reliable foundation for individual crop-growing and collective resilience to storms. Given the topography of Chikukwa and the cyclones that inevitably arrive periodically, they have had no alternative but to focus on water. After Cyclone Idai, CELUCT became the unifying organization that focused the efforts of the community's farmers to defend against such ferocious disasters in the future.

GUARDIAN OF THE TREES: GODFREY

On the first morning of our visit, we started downstream, visiting a farmer named Godfrey there. Godfrey told us about the different ways the community was coming together to respond to the cyclone by joining up individual efforts from the upstream sections across the entire valley. While this community effort started in relation to the disaster of the cyclone, the work is now hugely beneficial to the downstream farmers who manage the water from the rainy season so it is available for growing crops throughout the dry season. They have constructed interventions and ditches to catch the water so that instead of running off the land, the water sinks into the soil. Hundreds of cubic meters of water are held in the land through these structures. Godfrey stresses the need to hold the water upstream and slow its flow by building small, loose stone gabions to control its flow down the slopes. Now almost everyone in the village makes some kind of water-holding structures, which dramatically slow the flow of the water as it descends to the valley from the top of the mountain. Godfrey has an impressive nursery, selling tree seedlings of macadamias, bananas, coffee, avocados, and more. Farmers now have good harvests of yams, maize, herbs, and vegetables with the availability of ample water. The area has become a model for good practice at an individual as well as collective level, with shared benefits for all.

GUARDIAN OF THE CATCHMENT: SILAS

Our next visit, halfway up the mountain, is to a soft-spoken, gently smiling, unassuming 75-year-old farmer called Silas Chumure. The jeep climbs up along a steep, rough, rocky, bumpy road to the farm, feeling like it could overturn at any minute and send us on the fastest route back to farmer Godfrey! Silas, born in 1963, came to this inaccessible area 45 years ago. He chose to go as far up the mountain as he could, for the fresh air and the opportunity to work on any projects he wanted without being interfered with. He said he stopped where he felt free. Over the years since, in this freedom, Silas experimented with holding the water in many different ways, planting varieties of trees and looking after nature in his own way, using the rainy season water to recharge the underground supply. The water table is now raised, and a spring on the farm provides water freely all year round. When we asked him where he had learned to do such things, he replied that nature taught him how.

Here are a few ways that Silas tended the land around him:

- He experimented with plant propagation and trees. He took out the water-thirsty gum trees and instead planted water berry trees and wild fig trees, whose long roots helped bring the water from the water table to the surface. He knew that wherever he saw the wild fig tree, there would be water, as they grew only where the roots could access water.
- He moved his grazing animals around different patches of land so they didn't decimate a patch, allowing grass and saplings to regenerate.
- He took care of the upstream flow of water, largely on his own, building large numbers of staggered gabions along the river pathway and tight-fitting stone walls to slow the water down as it descended along the stream bed. His aim was to make the stream flow perennially again.

On a long, hot walk up his mountain, Silas took us alongside the forest and the wilderness to show us the 16 gabions, or stone walls, that

Gabion wall.

he had constructed over the years. With each gabion we came upon, we were surprised by the scale and elegance of the work that had gone into building these structures by hand in such difficult terrain and with only one set of hands! The sheer weight of one rock, let alone the hundreds needed to construct a single gabion, carried by one so slightly built as the wiry Mr. Chumure would have made a lesser man or woman give up at the very first gabion. Never mind going on for years until most of the stream was covered with the necessary water-holding structures, perennially watering and feeding not just his homestead but also those of a thousand people in the downstream village and valley below.

A gabion is a structure of stones, often held together by wire mesh or by tightly wedging the stones so that they interlock, used to stabilize streams from erosion and excessive flow. The 16 small structures strategically placed by Silas Chumure across the stream are made of loosely packed stones placed in the pathway of the stream flowing down his hillside, allowing small pools of water to collect behind them to temporarily hold back and slow the water descending the gulley of the streambed. Farmer Silas's gabions are cleverly engineered to slope toward the water so that they hold together naturally against the flow.

The gabions are each a work of art, perfectly positioned across the width of the stream, able to hold the water behind it, allowing the silt to deposit there instead of flowing on down to the valley floor. Some of the water held can seep into the ground, while the stones allow a slow flow of water to leak through and continue downstream. Silas Chumure has situated these stone structures so that they are secure enough to hold the volume of water that might fall from a cyclone, to protect the land further downstream.

As we climb higher the terrain becomes steeper and steeper, which gives us a new appreciation for Silas Chumure's determination, knowing that he has traversed this terrain umpteen times and carried each rock all this way on his own. We can only imagine where he found the rocks, how he carried them up the mountainside with each new gabion he built, and the sheer dedication and devotion needed to make his landscape and that of the village downstream water secure and healthy. He is truly a guardian of nature and the water within it.

By building the gabions in varying intervals along the valley from the top to the bottom of the catchment, Silas Chumure hopes that one day the water held there will refresh the water table enough to return the stream to gushing perennial flow.

Silas Chumure.

Silas has devoted his life to the area and oversees the restoration of the whole landscape above his farm. We decide together that Silas is the Guardian of the Catchment. He knows the land and the river and quietly works for their health. From his position on the catchment he is like a guardian who, knowing the young stream, can best understand its behavior and how to manage it as it matures along its course. We considered that an appointed guardian for each catchment could be a powerful idea to put in motion for future projects, ensuring that there is always someone looking at the health of the whole landscape they live in. The idea is not new and is found in many countries and traditions across the world. There is the tradition of "river keepers" in England, who were appointed to look after the health of their local rivers. When the position was dispensed with as an unnecessary expense, the rivers began to deteriorate—a lesson to us all.[3]

It could be said that a guardian like Silas Chumure is needed for every stream to ensure planetary water health. There is so much local knowledge, wisdom, and willingness that is untapped or disregarded in the chase for top-down control of water systems. If we want to manage the catchment, we must see the people who are working upstream as a critical part of ecological responsibility and governance of that waterway.

GUARDIANS OF THE SPIRIT: ANCESTORS AND FOREST SPIRITS

We say goodbye to Silas and go halfway up the next mountain, which is different from Silas Chumure's in that it is bare and decimated from the constant grazing of livestock. CELUCT is looking for sustainable solutions to the destructive pattern of overgrazing—without hurting cultural sensitivities—in order to get everyone engaged.

We arrive at the homestead of the youngish village headman, where a small gathering is sitting and, to our astonishment, already drinking beer at eleven in the morning! On being offered the beer, we are told that is a very special day for the area. To help usher in the rainy season, a harvest festival takes place, during which they consume the beer traditionally made from local grain grown by the farmers in the valley. They also offer this beer to their ancestors, in their special places of rest, and to the forest spirits so that the rains may come and herald the onset of a rainy season. This faith in their ancestral spirits has never failed them— it always rains within 24 hours of this beer offering. The community has been brewing this local beer for months in anticipation of the offerings that would need to be made on this day.

As we were visiting the water-holding work of the community in hot bright sunshine (and graciously being offered this beer), the thought did cross our minds that maybe this time they would not get the anticipated rains. The sun was just too strong and bright, with not a cloud in the sky.

GUARDIAN OF THE COMMUNITY: JOHN STRONG

John Strong on his land.

After this we found ourselves up the next mountain with farmer John Strong to see his farm. We had met John at the village headman's house partaking of the same festive beer. John, who is also in a position of leadership in the village and in some ways the voice of his community, gets his message across eloquently and clearly, emphasizing the importance of doing everything to get the water into the soil to recharge the water table. He leads us to some features on his own farm:

- Plantings of trees, including the water berry (*Syzygium cordatum*), Muonde wild fig (*Ficus sycomorous*), and a traditional variety of banana. The roots of these trees seek out the groundwater and become guides for the rainfall to enter through the cracks and passages into the aquifer. During dry periods, the roots work in reverse, drawing the water up into the leaves to evaporate and cool the surface.

- 72 community-built gabions upstream in the forest to slow the water descending to the valley.

- A spring on his land whose source is overseen by a magnificent Muonde tree. From the spring, a tangled collection of rubber pipes carry the water down to the main community buildings, the school, the clinic, and different homes, providing water to the community lower down. The CELUCT Center is also one of the beneficiaries.

John Strong then takes us through his fields even farther up the mountain, nearer the top. He has cows, chickens, and goats whose manure he puts on these fields. There are no chemicals used on his land. He has fencing to stop the animals getting into the fields. He shows us a field prepared for coffee. Since 2005, he has planted many varieties of trees: bananas, mulberries, fig trees, lemons, plums, macadamia trees, and even peaches. He has also planted water-absorbing vetiver grass, along with the trees, to stop the water flowing down the slope and eroding the soil. The fields are filled with the most magnificently healthy maize, cabbages, beans, and other vegetables; even the small spaces in the gigantic rocks have been pressed into growing fabulous cabbages!

Lengthy pipes taking the water from the spring upstream to the households downstream.

Before John came with his water-holding program, the rain used to wash down the slope, carrying all the soil and flooding the fields beneath. But now the water is properly held so that it enters the soil, recharging the aquifer, and there are bumper harvests every season.

Cabbages growing between the rocks. Courtesy of The Flow Partnership.

HIGH PASTURE REGENERATION

After talking with John Strong, we go even higher—up to the very top of that mountain. Sam of CELUCT takes us to view how concentrating the grazing of cattle has encouraged the return of the grasses, which in turn allows the water to sink into the soil.

Sam has introduced a system of managed grazing to the area. When animals were domesticated, humans started protecting their sheep, goats, and cattle from predators so they could spend many days contentedly grazing in the fields. This also meant that they ate from the same fields every day, so their continual grazing stressed even the root system of the grasses until the whole plant would die. The land would become ragged and bare.

Instead of allowing the cows, sheep, and goats to roam wherever they wanted, grazing in the CELUCT area is carefully managed. The village animals are taken through a strategic grazing plan, where they graze the grass in one area for a short period before being moved on to the next through fencing. This short-term grazing pattern allows for the recovery of the grass and its roots, aided by the vast amount of droppings left behind.

The mountainside still showing the scars of landslides from Cyclone Idai and bare land from overgrazing where new practices have not yet reached. Courtesy of The Flow Partnership.

This practice originated with herds of animals bunching together to keep predators away. They would migrate from place to place before returning to any one spot, to keep other animals off their track. The disturbance of the soil and the provision of manure benefit the grasses and the land. As a result, the grass and young tree shoots were able to thrive, regenerating the land, improving the soil, and improving water infiltration.

In Chimanimani, the sloping surfaces sent heavy rain as runoff to accumulate as floods in the valleys. In places with managed grazing, grass has been allowed to regenerate so it can hold back the rain and help the water infiltrate the soil. Instead of water falling on a hard surface and running off, the grass recovery raises the water table of that piece of land over time. The water is retained, the mountainside is able to regenerate, and landslides are prevented by binding the soil through the roots.

Throughout the area, there was a marked difference to be seen. Where the animals grazed freely, the land was bare and the soil had hardened into a crust. When rain would come, it would just run off, as from concrete. Where the grazing was managed, the grass softened the soil and allowed the rain to infiltrate into the ground. Sam showed us places where this practice of maintaining the grass alone had resulted in streams reviving to provide households with water.

The panorama from the top of the mountain is magnificent, with views stretching to the Chimanimani National Park and on across the mountains to Mozambique. We felt a great sense of unity, as if we had been granted a secret view into nature's wisdom. When various community members do the most basic things right, their water-holding work can change the very nature of the crops and the health of the valley. The lessons learned from the deluge of Cyclone Idai are visible for all to see as the community puts in the effort to prepare for if a similar cyclone comes again.

PATHWAYS TO CHANGE

Back at CELUCT we talk with Mybe Patience, one of the original founders of the project. In the last 30 years, the area has seen less runoff, the soil has improved, and the trees have grown. Chikukwa consists of seven villages, with one chief and a council of elders. The elders uphold traditional ways of working with the land and also spearhead new ideas, one of which was setting up CELUCT.[4] The chief helped the community systematically improve the whole area through holding water and planting more water-loving trees. The elders of the community understood that they were carrying on the work of their ancestors by looking after the land. Mybe told us that the women were often the first to experiment with and adopt new sound ecological practices on the land to improve their capacity of supporting life. Sometimes the men would be working in towns away from Chikukwa, and they would see the changes made by the women only when they returned home on weekends. Through the workshops held by CELUCT, the social structure also changed: people learned from each other in discussions that involved both men and women. Men realized that the women, when empowered, were the ones who could change the land. The women could now decide what they wanted to plant in their fields without needing to get permission from the men.

One of the primary projects of CELUCT is to help the farmers receive basic training about holding water and uniting their efforts around permaculture as a way of life. Water harvesting and soil conservation feature heavily in all the cross-community trainings. They encourage the community leaders to come together, to learn together, and to implement together. They assist the farmers with planting indigenous flora along the stream and give the farmers fencing material to protect the young shoots of trees and grasses from being overgrazed or destroyed by livestock and animals.

CELUCT has also developed a conflict transformation program that builds peace and community cohesion to deal with conflicts related to unequal access to natural resources (not enough sharing of water, land, and the like), power struggles, marginalization of women and children, and youth whose future has been impacted by unfair access to relief food and inputs.

Before this restoration work, it was mainly the elders living here, with many young people having migrated to the larger towns and cities. Now, with the return of community cohesion and rich subsistence farming, the young people are happy to grow up and stay here again. Dalend, the head of CELUCT, grew up in Chikukwa. He remembers as a young man taking the goats to the pastureland at the top of the mountain, a job he used to love and which he now sees youngsters committing to again. Now he presides over the valley with a new responsibility, not just preparing to withstand the next cyclone but also—unlike in some other parts of the country—to live in peace and harmony in an area abounding in rich, natural, healthy food and water, with pleased ancestors blessing them with a rainy season arriving on time.

By about 10 p.m. that same night, the clouds gathered in depth and intensity, with lightning and thunder like we had never seen or heard before. And then, well before midnight, the heavens opened up, the rains came, and so much water fell from the sky that we were worried the road out of the village would become impassable again the next day. The ancestral spirits had ensured the villagers' hard work paid off and the rainy season had arrived on time! We were heartened to see this complete harmony between the villagers and their ancestors—the communication channels seemed to be functioning perfectly.

Burkina Faso: Transforming the Desert

Six thousand years ago, the Sahara was a savannah of trees, shrubs, and lakes—full of life.[1] Today, one can traverse the hundreds of kilometers of the Sahara between northern and central Africa and find little but hot, sandy desert. On a given day, flowers might bloom spontaneously following exceptional rainfall or the odd cactus might drop its gourd-like shell on the ground as a single mark of lifelessness. The animals, whose fossilized impressions still lie in profusion beneath the surface, the forests, the lakes, the human habitation are all long gone from this sandy silence. As far as the eye can see, desolation and lifelessness stretch endlessly.

THREE PATHS TO DESERTIFICATION

There are diverse theories about what happened. Of course, nothing definite can be known about how such a shocking and extreme transformation happened where the European, African, and Asian landmasses meet. This transformation from dense forest to dry sand indicates that the water cycle is not a permanent system that stays in place forever once a climate pattern is established. The water cycle develops dynamically and reflects the framework of the changes within which it operates.

To understand what happened to cause that shift in the weather and the water cycle to go into a seemingly irreversible breakdown, we have to look back 6,000 years. We looked at three theories, among others, for this shift in the Sahara from savannah to desert:

1. As humans and domesticated animals learned to live and graze for long periods on the same area of ground, the shoots of new trees were systematically killed off. Some suggest that this practice of gradually cutting the connection of the groundwater aquifer to the surface, enabled by tree roots, contributed to the breakdown of the water cycle.[2] Did humankind, as it set out on a pastoral existence, take too much liberty in its efforts to control the land, farm crops, and domesticate animals? Did that disrupt the balance of nature and prevent the rains from reaching the interior of the Sahara?

2. There is evidence that an "orbital wobble" of the earth, due to its gravitational relationship to other planets, subtly changed the orientation of the earth to the sun, and so disturbed the seasonal currents bringing rain to the area.[3] The earth's tilt weakens monsoon currents by diminishing the amount of the sun's energy powering their course. Was this tilting of the earth's axis a contributing factor to the desertification?

3. Once the Sahara started becoming bare, there were fewer and fewer trees to hold and circulate the groundwater, leading to the climate drying out. We conjecture that a third decisive shift happened then. As the Sahara became a heating rather than a cooling source of winds, these hot winds would have pushed rain away from the area. Different factors can tip the climate suddenly into quite a different pattern—here we see desertification as the effect.

While we will never definitively know what happened to the Sahara 6,000 years ago, we can observe and experiment with what is happening today in its borderland, known as the Sahel, to keep the desert at bay. The questions being asked are, can we shift the current pattern of rapid desertification in the regions comprising the Sahel neighboring the Sahara? If so, how?

Uncomplicated, on-the-ground work with communities caring for the land, holding water, and planting trees systematically reverses the desertification process, turning arid landscapes back into lush greenery. By changing the practices in the simplest ways, such as letting old trees (or stumps) regenerate, keeping the animals away from the tree shoots, and using simple methods to hold the water in the fields, the landscape of the Sahel can be made to flourish again. But for it to flourish and become green again, this will have to be done at scale.

THREE ROUTES TO TRANSFORMATION

We look at the work of three successful organizations working at the boundary of the desert and returning a flourishing agricultural environment to the Sahel:

- Groundswell International, scaling up individual practices for a global impact
- Farmer Managed Natural Regeneration (FMNR) community regeneration of forests
- Terre Verte, reviving the water and soil, field by field

Groundswell International

Groundswell International, one of the partners of the Water School Africa, is working to coordinate projects happening at an individual scale into a larger scale revival of the area. Tsuamba Borgou, the coordinator of the West Africa Network of Groundswell International, now in his 50s, tells us that when he was a child growing up in Burkina Faso, the area he lived in was still green and abundant with many trees. At the time, what felt remarkable were the small patches of land where the vegetation had dried up or the trees had been cut. But now, a few decades later, the Sahara desert has crept further and further into this territory and what catches the attention is the opposite: small patches of trees or islands of vegetation that stand out within the aridity of the landscape. The land has become a drought-prone, semidesert area. A rainfall of 500 millimeters has become the baseline when 900 millimeters used to be quite common. When he was a child, the rainy season lasted from May to November; now the rains are restricted to falling between June and September. After that, there is not a cloud in sight for the rest of the year.

This pattern of dramatic change in the landscape that Tsuamba describes is a typical consequence of the water cycle not working any more. The Sahara, in its desert condition and holding extreme heat, resists the seasonal advent of the African monsoon rains. More rain might fall sporadically farther south in Africa, but the pattern of rainfall slowly disappearing from the Sahel is intensifying the pressure on the scant vegetation left. The more the land heats up, the further resistance there is to the atmospheric currents bearing rain. A loop of desertification plays endlessly.

Some of the methods the Sahel farmers use to hold the rainwater and enable it to sink into the ground are:

- The zai pit, which is a small hollow that can be dug by hand, into which compost is placed and a crop or tree individually planted.
- A half-moon structure that creates a crescent-shaped pond whose bank holds the runoff water flowing into it.
- *Cordons pierreux* (stony cordons), which are thin lines of fist-sized stones that form a barrier to slow the flow of water across a field and trap sediment behind them.
- Sand dams, which are check dam–like structures where the water is held and protected from evaporation below the silt and the sand.
- Trenches that can be dug in the ground so that the water flows slowly and stays as long as possible on the farms when it rains.
- Bunds—small individual check dam–like structures built above ground that halt the water flowing across the land and allow it to stay in each field.

The consequences of concentrating every drop of water with any of these methods to feed into the cropping areas are profound. The arid landscape, where the winter rains would simply run off the surface and disperse, transforms into a lush, green land, where it seems almost like each plant and tree is given the individual attention needed for it to reach maturity. Once given the conditions needed to return, life finds its own unfolding through a small forest, a lush maize crop, beetles, and birds.

Groundswell International has a successful history of connecting grassroots organizations and applying local solutions to resolve a global challenge at scale. The Green Revolution formula of chemical fertilizers plus irrigation equals production cannot override the unmatched combination of soil, organic matter from plants, and water as the dynamic foundation of life. It is no wonder that the earth has been left as dead ground when we forget this. When farmers attune again to the language of water, they relearn the natural formula for life.

Lessons from the Great Green Wall Initiative

An ambitious multibillion UN-sponsored Great Green Wall Initiative was started in 2007, committed to turning back or halting the desertification of the Sahel. Over $14 billion were newly promised in 2021 to plant billions of trees in a 30-kilometer wall at the edge of the desert and prevent its further spread.[4]

The UN report on this project, however, catalogues a large-scale disappointment:

> Stakeholders interviewed have reported that, when the African Union set up the PAA (Pan African Agency of the Great Green Wall), there were disagreements over basic issues such as staffing and posts needed. It seems that consideration was mainly given to senior appointees with political credibility, irrespective of the nature of work to be done. The agency therefore appears to be "top heavy," with an inadequate structure below this to ensure that basic work is completed in good time and quality.[5]

Another section of the report considers the Sahel and West Africa Program supported by the World Bank. Here is the report's evaluation of on-the-ground work:

- primitive data-monitoring techniques
- lack of evidence of claimed changes
- untargeted programs with mixed objectives beyond tree planting
- fluctuation of reported dense forest levels rising and falling year by year, which illustrate their inaccuracy
- when trees were planted, it was implemented in an overly expensive way, mainly in Ethiopia[6]

The report reflects only sporadic on-the-ground progress in actually building the Great Green Wall (GGW) because of the tangle of political complexity. What was forgotten in all the promises and speeches made when it was launched was that a tree or plant is not simply the figment of an investor's imagination. A systematic knitting together of the local communities is required to achieve success at that scale. All those good words washed away with the first rain and then were baked hard into the unremitting ground of a soil that no one had taken the trouble to awaken. The UN review continues by analyzing the reasons for the ailing initiative:

> Both international and country stakeholders ultimately attribute the limitations on performance of the PAA to weak political support. The Council of Ministers of the GGW is said to meet irregularly and to have failed to ensure resource viability of the PAA Secretariat. In turn, the African Union, to which the GGW reports, is seen to be lukewarm in its support, failing to elevate the Initiative from the status of an ambitious vision to that of an implementable program. The overwhelming consensus among stakeholders is that the functions and approaches of the Accelerator cannot be successfully transferred to the PAA in 2025, since the Agency is in no position to absorb them. The Review finds the current stakeholder perceptions challenging for the future of the GGW, since it is difficult to find anyone who sees this transfer as feasible.[7]

The failures of the Great Green Wall Initiative bolster the case made in this chapter that regeneration of landscapes needs to be in the hands of the communities that live in them from the very start. It is in the interest of the communities to regenerate their area—they are the ones who know what works and what doesn't. They know that first you have to hold the rainwater in the ground and not let a single drop escape. This recharges the groundwater. Only then can you plant trees and halt the desert. Any resources that can encourage the communities to do that will be far more effective and successful at a fraction of the cost.

Tsuamba had the additional insight that it not enough to enable one field to produce crops for an extra season or enable a small woodland to arise. Although these are important steps, the real transformation of the landscape, the reversal of the land's desertification and the return of a balanced climate, can only happen when the water is intentionally held over entire catchments. This will really make a dent in reversing the encroachment of the desert.

In a recorded interview with Tsuamba Borgou, he summarizes:

> One farmer alone cannot raise the water table. It needs collective action of the community along a whole river catchment to do this. Where it rains a lot, there is an indifference in modern communities to the notion of the river. But where it is dry, then the experiment of how to revive the water can be contagious. By training the farmer in methods to hold the water, the farmer is also able to tell the neighbor, so the water challenge is taken up by the whole community.

Can the local community methods of holding water in the ground be scaled up enough to change the behavior of a whole catchment or a region? Tsuamba continues:

> The "new" idea is water table recharging. The previous work may have contributed to that but it was not intentional. The intent was to grow crops and increase yields. The need is to base the project on what is being done at very small scales and weave these multiple small works and make them more systematic. It requires more training to make the water recharge intentional and to show that to have a large impact, one needs to do it at a large scale. That is what I think we should concentrate on.[8]

Paying attention to water going into the ground will create the combined action of raising the water table around the whole area. When you knit the individual farmers into a whole catchment and whole catchments get linked through replenished underground water tables, then the region has the wherewithal to change. We need a local spread of water literacy and action in an increasing loop if we want to regenerate and revive entire large regions and eventually tackle a land mass the size of the Sahara. A Great Green Wall can be built if done with love and care, nurturing the water and life in the region, systematically and with attention to ownership and collaboration with the local communities living along the length of that proposed green wall.

Farmer Managed Natural Regeneration (FMNR)

An example of this community-led revival is the success of Farmer Managed Natural Regeneration (FMNR),[9] which is enabling impactful work at scale. In contrast to the unsuccessful Great Green Wall reforestation project in the Sahel or the practice of bringing exotic foreign species into the landscape, as was the standard development method of reforestation in Niger before the 1980s,[10] FMNR enables the regrowth and expansion of what is already there. The FMNR program started in 1983 with 10 farmers in Niger experimenting with the underground yet living and sprouting tree stumps of indigenous vegetation. Today FMNR is spread across Africa and many other countries in the world.

The FMNR approach seeks to empower farmers to regenerate the forest itself by managing, protecting, and nurturing the many seedlings from the existing trees that are on the ground. Shoots from living tree stumps or roots in fields, pasture, or woodland appear as small shrubs. For many farmers these shrubs are in the way of their crops and need to be removed or are used as general feed for the grazing animals. FMNR recognizes that the tending of shrubs into trees can coexist with the growing of crops and the grazing of animals. After recognizing living stumps and deciding how many regrowth shoots to allow, some of the weaker shoots will be harvested for wood, while the stronger shoots will be allowed to grow into trees. Cutting and selling the excess young trees for firewood and timber provides valuable revenue, while leaving the more hardy shoots to develop into trees.

The process has multiple benefits at once:

- better uptake of water through the trees' roots
- reclamation of degraded land
- increased crop yields and animal productivity
- increase in biodiversity
- transformed quality of life in the villages

- the availability of fire and building timber for use and for sale
- the spread of these methods based on their success[11]

The success of the FMNR program is that the farmer who owns the trees benefits immediately from the work he does to tend to the saplings growing from the living tree stumps. Instead of imposing a new regime on the landscape by importing new species for wood production, the process recognized that the Sahel still had the possibility of life within it, and this life could be nurtured back into abundant maturity through what was already there.

The individual farmers get clarity about their role in a process that can transform the whole landscape from semiaridity into life. They focus on providing the individual plants with a self-contained ecosystem in which they can grow. Restoring the indigenous water-loving trees helps make the groundwater accessible through the roots and aids the work of the water cycle. The tree is not just an isolated thing growing in the landscape. It is also adept at holding the wet season rains through the soil and groundwater and then accessing that groundwater during the dry season. While pictures of regeneration concentrate on the renewed greenery in an area, unseen in the work is the restoration of the whole water cycle, through the roots, through holding water in the ground, through the air that is cooled by evapotranspiration from the leaves, and through the conditions made more amenable to receiving the African monsoon rains.

In 1984, in response to a famine in Niger, a food-for-work program was introduced (by Serving in Mission[12]), training some 70,000 people on 12,500 hectares of land in FMNR techniques. By 2001, it was estimated that "20 million trees had been naturally regenerated and pruned for multiple use in the Maraudi region [Niger] over the last 17 years."[13] By 2004, 50 percent of Niger's farmland, or 250 hectares annually, was practicing FMNR. The practice then spread over all of Africa, including in Burkina Faso.[14] Now the number of trees in Africa as a result of this work is so numerous, it is beyond counting.

Terre Verte

Terre Verte is adept at changing a landscape system one farmer at a time. They have been working in the Sahel since 1989, creating a stakeholder group to devise a common strategy of water harvesting and land restoration on individual plots and collective areas. It took 15 years for Terre Verte to work out a sustainable model of farming, beginning with their first pilot in 1989 at Guie.

Terre Verte has introduced the following:

1. The co-owned common area is protected with a perimeter; it contains an enclosure for keeping animals in during the rainy season and includes ponds for holding water from which animals can drink.[15] The common area is divided into individual plots to grow trees and crops at a farmer level, and each individual plot contains a number of fields.

2. Shifting attention away from the productivity of what is growing, the perimeter of the whole site containing the individual fields is cultivated to clear a fire break around the fields. This acts as a barrier to stop animals entering and offer securely gated entries to allow necessary access.

3. The zero-runoff technique that focuses on the conservation of rainwater and soil throughout the plot means that no drop of water leaves the field through runoff: the water infiltrates into the soil and can only leave the field through evapotranspiration from the soil and plants. This is mainly achieved by creating zai pits, earthen bunds, hedges, and ponds.

The collective area is divided into plots of between two and four hectares, managed by individual famers. The plots are further divided into three or four fields. An informal co-ownership highlights the individual's responsibility within the collective need. The plots are made up of square fields that are set up around a bund placed according to the slope to keep all the water that falls within the field. This is a pioneering water-holding technique that focuses on each individual field for holding the rainwater in flat lands. A field that is, at maximum, 50 meters wide is able to manage torrential rains without erosion.

The fields are surrounded by an earthen bund lined with a hedge, with a pond at the lowest point to hold and infiltrate excess water from runoff. Large trees are planted on the edges of the field so they do not interfere with the cultivation of the field.

An update from the Terre Verte website tells us the experiment has led to a solid foundation for a prosperous farmer: "Farmers then have an excellent working environment, ensuring good yields and being sustainably productive."[16]

Tending the Perimeter of a Field

The life of the field is not just about the productivity of what is grown in the field. The success of the field begins with the perimeter, which decides what comes in and what stays out. The field perimeter is carefully designed to protect the relation to the outside. This consists of three layers:

1. a cleared firebreak outside the perimeter, which protects the field from the risk of fire
2. two lines of shrubs around a sheep fence, which prevent animals from entering the field
3. secure openings for entry and exit into the fields

In this way the field becomes its own unit, secure from animals, with its own bund to stop runoff and a pond to hold the water. The field system optimally enhances the water cycle to reestablish greenery. A path connects all the fields, so the individual farmer is also placed within the collective effort.

ZAI PITS AND ZERO RUNOFF

Zai pits concentrate water and nutrients around a plant or tree in Burkina Faso. A turning point in the Terre Verte experiment was the discovery of zai pits in 2006 as a means of growing individual plants. The lesson of the zai pit is in its independence. Each tree or plant is individually given the nutrients and water it needs.

The zai pits can be productive for crops from the first year, gradually restoring the degraded land. They also avoid the need for ploughing, which often breaks up fragile soil and risks it blowing away. For smaller plants, the holes are dug 30 centimeters in diameter, 15 to 20 centimeters deep, and 70 to 80 centimeters apart. The excavated soil can be formed into a ridge downslope to help the rainwater stay in the hole. When planting trees, the zai pits are larger, deeper, and placed farther apart.

Zai pits are suitable for flat or gently sloping land with low rainfall of between 350 and 600 millimeters annually. A heavy cultivator tool can be used to safely open the earth, vibrate the soil loose, and allow good infiltration of rainwater.

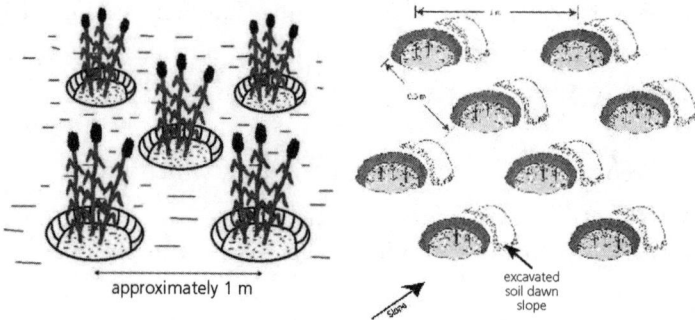

approximately 1 m

excavated
soil dawn
slope

Zaï pits (from Burkina Faso). Mati, B. M. 2005. Overview of water and soil nutrient management under smallholder rainfed agriculture in East Africa. Working Paper 105. International Water Management Institute (IWMI), Colombo, Sri Lanka. Published in Mekdaschi Studer, R. and Liniger, H. 2013. Water Harvesting: Guidelines to Good Practice. Centre for Development and Environment (CDE), Bern; Rainwater Harvesting Implementation Network (RAIN), Amsterdam; MetaMeta, Wageningen; The International Fund for Agricultural Development (IFAD), Rome. (Courtesy of Wocat)

In the case of the Terre Verte plots, zai pits are sometimes supplemented with *cordons pierreux*, which are lines of stones that hold the water runoff and increase the chances of the water entering the zai pits. Rain gardens near homes and large ponds that hold and store the water on flat land and bunds also ensure that all water that falls stays in the field. The individual plots form a common effort that improves water storage, transforms the landscape, develops agriculture, and raises livestock.

The individual homes and families must also commit to:

- maintaining the earthen bunds so that they are not eroded with heavy rains
- maintaining the common areas (paths, firebreaks, mixed hedges, ponds)
- trimming the hedges so they keep their shape as the perimeter boundary

THE MAN WHO STOPPED THE DESERT: FARMER YACOUBA SAWADOGO

Traditionally zai pits were only an occasional method used to help re-green barren land, but their use became regular and widespread after the phenomenal results of the farmer and agronomist Yacouba Sawadogo from Burkina Faso, whose inspiring story is well told in the film *The Man Who Stopped the Desert*.[17]

Sawadogo brought in two main advances in the use of the zai pit. First, he extended the size of the hole to be big enough for a tree. He also dug the hole earlier, while still in the dry season (as opposed to waiting for rainfall). The zai pit was filled with manure, rocks, and a small quantity of soil and seeds of cereal (millet, sorghum, or corn). The manure attracted termites, whose tunnels helped to break up the soil further and whose droppings were more readily assimilable by the growing plants.[18] Research on the impact of zai pits shows that yields of sorghum are routinely raised from 500 kilograms to 1,500 kilograms per hectare.[19]

Sawadogo worked with zai pits to regenerate the soil and plant trees, creating a forested area of 62 acres, or 250,000 square meters, out of the desert. Surprisingly, after the forest (called Bangr-Raaga, which means Forest of Wisdom) was generated by Sawadogo, the ownership of this forest and land was disputed by the government. Ironically, once the land became green and fertile, it attracted land investors, who started cutting down the newly grown trees to make way for housing in that regenerated area. But in 2021, after Sawodogo had won a number of international awards for his work, the area was sealed off within a protective fence by the minister of the environment. This fence seems to have protected the forest as municipal heritage and rescued the land from the housing developers for the time being.

Zai pits overseen by Yacouba Sawadogo. Image from *The Man Who Stopped the Desert*.

THREE STEPS TO REVIVAL

In summary, we can now make the bold claim that it is possible to reverse some of the reasons why the Sahara savannah turned to desert 6,000 years ago, using the steps communities have been taking to transform the aridity of the Sahel back into forest and life. The three theories we posited at the beginning of this chapter now stand as follows:

1. Holding water, protecting shoots from grazing, and allowing indigenous trees to spread, practices used by Groundswell, FMNR, and Terre Verte, replenish the groundwater in the rainy season. This water is drawn up by the tree roots during the dry period. In the work we have documented, holding the water, planting trees, and nurturing a forest is a natural pathway to the restoration of the water cycle and the transformation of arid lands to fertile fields.

2. The effect of the "orbital wobble" of the earth is a periodic occurrence affecting whether the African monsoons bring seasonal rains to the Sahara. This orbital wobble is going to recur, pushing the dry period into a future wet season (in so many thousand years).[20] We would do well to prepare for this by having structures in place that can hold the water when it falls and forests that can channel the rain into and out of the aquifer and the water cycle.

3. Every neighboring African country suffers from the hot, muggy, debilitating, sand-carrying wind that blows from the Sahara. Each country has a word for it, so that newcomers can be alerted to its tribulations and tendency to foster bad moods. This wind that bodes no good is also perhaps what finally closes the door on the African monsoon rain. This aggressive self-generating heat barrage repels the life-giving currents of rain. It is vital to work toward that tipping point where the Sahara can be contained until the monsoon rain can push open the door again for life to flourish.

How the planet regenerates is in our hands, whether from savannah to desert as in the Sahara 6,000 years ago or from aridity to fertile fields, flourishing communities, and reliable rainfall in the Sahel. The work of holding water in different ways across different terrains and returning greenery, trees, and forests can cool the earth and allow rain-laden winds to reliably foster life. This is valuable for its own sake at a local level, but also, when done at scale, this work can enable the planet's climate to stabilize and create a good life on earth for generations to come.

10

Slovakia: Building a Rain Garden

As climate change and extreme weather events become the defining issues of our time, we all want to do something to help resolve these complex and global issues to continue living on a healthy planet. Many city dwellers, living in apartments and city blocks with little personal land available to them, wonder what they can do about these issues at a local level. One method for many people with access to a small patch of land is to create a rain garden.

WHAT IS A RAIN GARDEN?

A rain garden is an attractive gully in a garden used for capturing rainwater from sealed and impermeable surfaces such as roofs, sidewalks, roads, and parking lots.

At the Summer School of Water 2019, organized by People and Water International,[1] held in Kosice, the second largest city in Slovakia, Peter Bujnak helped us create two rain gardens for a school and taught us a simple lesson: a rain garden is very similar to a normal garden. The main difference is that instead of using water from the tap to irrigate the plants, a rain garden uses diverted rainwater runoff—much more efficient and effective.

Peter Bunjak states: "In the current climate change debate, a rain garden supports two key features in a functioning water cycle. One is evapotranspiration, where the collected rainwater needs to evaporate, refueling the small water cycle. The second is infiltration of water into the ground."[2]

In the cities, tens of thousands of cubic meters of rainwater run off each roof, draining away through the gutters into the sewers, sometimes into canalized rivers and sometimes out to sea, often contributing to floods, droughts, and climate change. The concrete surfaces of the city and buildings stop the water from entering the soil below, drying it out and breaking the cities' water cycles, making them even hotter. This is where the city and town dwellers come in, ready to unite around creating local rain gardens.

The principles of building a rain garden apply to the design of any water-holding intervention in the landscape: slow, store, and filter water into the ground.

WHY RAIN GARDENS?

A rain garden can capture rainwater running off from covered or sealed areas such as roofs, pavements, parking lots, or roads. Doing this helps

- re-energize and replenish the groundwater in the city
- add to the attractiveness of city and municipal gardens
- improve the water quality through soil filtration before it enters the local watercourse
- neutralize flooding
- green the street, town, and city
- aid the microclimate through increased evaporation
- reduce rainwater drainage costs
- strengthen the water cycle
- employ the simple principles of water transformation
- nurture a sanctuary for wildlife, including birds and butterflies
- cool the city so it doesn't become a heat island

Wherever we live, we can all help enable rainwater to be held where it falls. If all homeowners employed measures to do this, much of the rainfall could be restored to the local small water cycle. Anyone can build rain gardens. They can be sized according to available space, which, in addition to cooling cities and helping reduce floods, also provides a green and pleasant environment.

A rain garden (left) versus a drain (right). Courtesy of Peter Bujnak.

Michal Kravčík, the founder of People and Water in Slovakia,[3] has done a significant amount of work with the Slovakian government to bring multiple interventions to hold the water and stop flooding. He has been able to frame this local work to restore the water cycle very powerfully and clearly in his book, *Water for the Recovery of the Climate: The New Water Paradigm*,[4] which contains a wise understanding of the planetary water system—it is well worth paying attention to the lessons therein.

Kravčík notes that land began gradually to dry up after the Industrial Revolution of the 19th century, when soil was sealed under impermeable surfaces, trees were cut down, and forests were transformed into agricultural landscapes and then to urban landscapes, all of which cause the rain to run off the land, draining it out of the ecosystem. This is all water lost from the local water cycle. This amounts to roughly 730 cubic kilometers of water lost throughout the world per year.[5] We need to change the makeup of the landscape to allow it to retain water everywhere. Rain gardens allow all citizens to work with nature and participate in managing water in our towns and cities so they can become better places to live and work in.[6]

The question people always ask is "Do we need to have land in order to build a rain garden?" Often the answer is no. We can build rain gardens in small spaces that are being used for other purposes (such as parking lots, cement verges outside houses or apartment blocks, or traffic islands), and make them permeable surfaces that allow rainwater to soak and filter into the ground instead of letting it flow away via the drainage system into the sea, creating beautiful and functional spaces in the city with the same action. What's not to like if a concrete parking lot is now not only a place to park cars but also has beautiful, flower-filled rain gardens along its edges that can help cool the local temperature and give a natural boost to the local small water cycle from this captured rainwater?

Danka Kravčíková, the organizer of the rain garden dig as part of the Summer School and the cofounder of Water Holistic, along with her father, Michal Kravčík, reiterates that holding rainwater in the soil is nowhere else more critical than in cities. The draining of water leads to a harsh, hot, mainly concretized environment that often becomes too hot for any life, especially now that the planet is heating up at an accelerated rate. In addition, after extreme storm events, such surfaces speed the runoff of the water to a point of vulnerability: a flash flood in the city, an overflowing dam downstream, a low bridge collapsing, or a narrow channel overflowing, unleashing billions of dollars' worth of damage, not to mention the loss of lives and untold misery. Water runoff can also make cities so dry that governments have to resort to trucking in drinking water for the residents from hundreds of miles away, which can often cause shortages in the places where the water is being taken from.

BUILDING A RAIN GARDEN

In Slovakia, we were able to create two rain gardens for a local school with the collaboration of the local mayor, the administrative authorities of Kosice, the school staff, and the local community of parents and children, who spent their weekends working with us on the project. The location was particularly important as it was right in the heart of the city, serving as an example for the entire neighborhood as well as the rest of the city. During the two days it took to create the rain gardens, neighborhood residents kept dropping by asking questions and wondering how they could create rain gardens for their own building complexes.

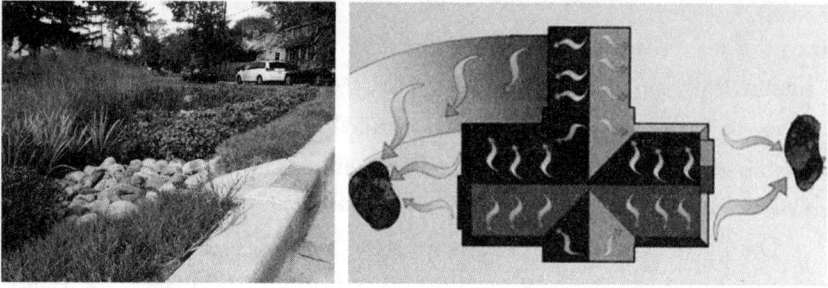

Rain garden (left) and outlet from roof (right). Courtesy of Peter Bujnak.

Step 1: Calculate the required size of the holding structure. The first step is working out how much water storage is possible for the space—only then can you work out where to situate it.

Three factors help determine the size of a rain garden:

- the volume of rainwater runoff from roofs, driveways, and other sealed surfaces
- the depth available for the rain garden
- the type of soil in the space

Step 1.1: Determine the size of the garden based on the area of the sealed surfaces. First we measure the runoff area of the roof from which water will be collected into the structure. In our case, the school roofs are 30 square meters each. As a rough rule of thumb, the area of the pond should be created by taking the area of the roof and using a ratio of 1:5 for permeable surfaces such as sandy soils or 1:3 for less permeable soils such as clay. In this case, with a clay soil, the 30-square-meter runoff area is multiplied by 1:3 to determine the raingarden's size at 10 square meters.

Step 1.2: Determine the depth of the garden. The soil should be able to drain the water that falls quickly, at least 1.25 centimeters per hour. If it takes longer because of heavy clay or a high water table, then the site is not suitable. This can be tested by digging a hole 25 centimeters deep, filling it with water, and observing how long it takes to drain. Between 1.25 and 5 centimeters per hour is a suitable draining rate for

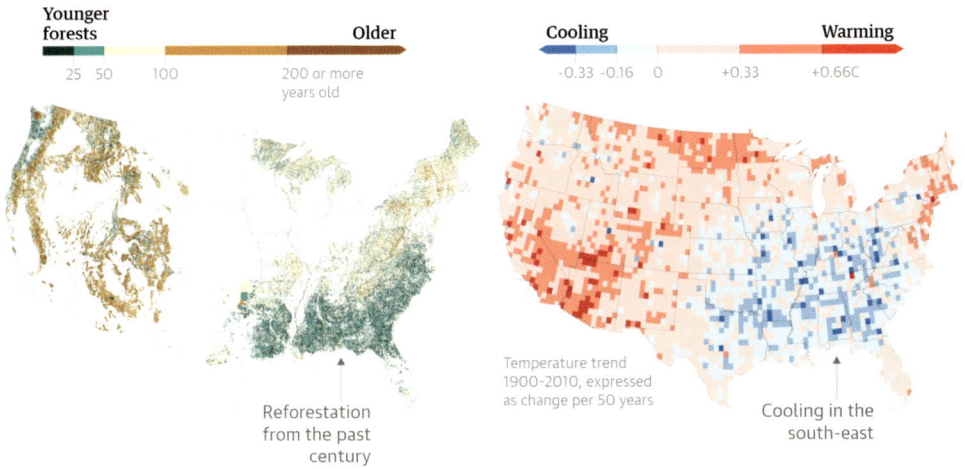

Figure 1. Eastern and Western US reforestation and temperature correlation. Reproduced under license from the Guardian article Milman, O (2024) Very cool: trees stalling effects of global heating in eastern US, study finds; 17th Feb 2024 accessed online on 17th May 2024 on https://www.theguardian.com /environment/2024/feb/17/us-east-trees-warming-hole-study-climate-crisis

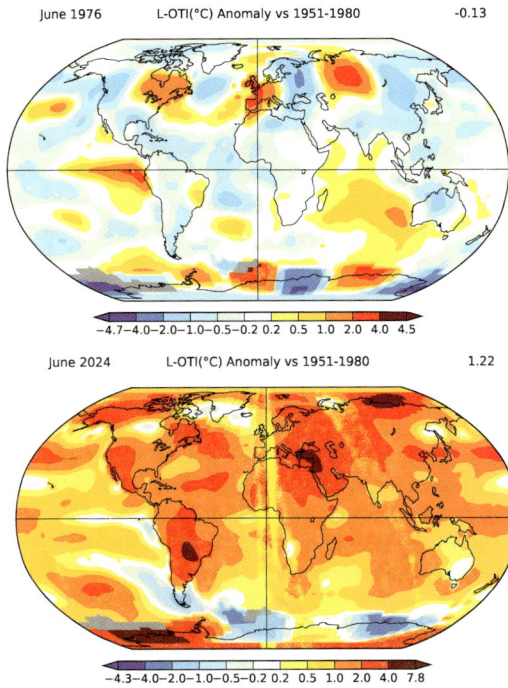

Figure 2. NASA GISS Surface Temperature Analysis (v.4), Global heat anomaly maps from April 1976 and April 2024, accessed May 30, 2024, https://data.giss.nasa.gov/gistemp/maps/.

Last 9 Years Warmest on Record

Global Temperature Anomaly (°C compared to the 1951-1980 average)

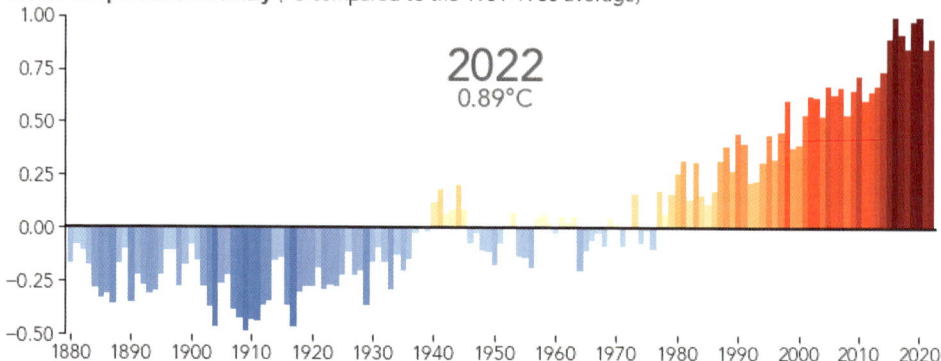

Figure 3. Increasing global temperatures as annual average. Courtesy of Earth Observatory, NASA Earth Observatory, 2023, "Last 9 Warmest Years on Record," accessed May 30, 2023, at https://earthobservatory.nasa.gov/world-of-change/global-temperatures.

Freshwater withdrawals as a share of internal resources, 2020

Freshwater withdrawals refer to total water withdrawals from agriculture, industry and municipal/domestic uses. Withdrawals can exceed 100% of total renewable resources where extraction from non-renewable aquifers or desalination plants is considerable.

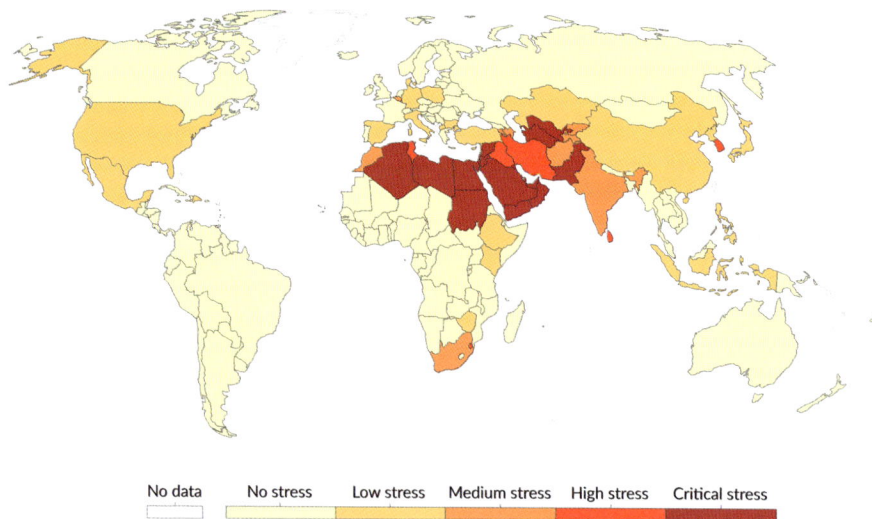

Data source: Food and Agriculture Organization of the United Nations

OurWorldInData.org/water-use-stress | CC BY

Figure 4. Water Stress. Courtesy of Our World in Data, 2023, "Freshwater Withdrawals as a share of internal resources," 2019. Accessed May 30, 2023, https://ourworldindata.org/water-use-stress.

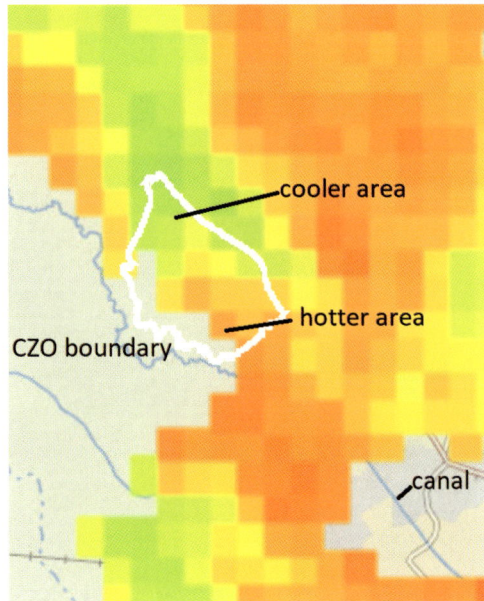

Layer of satellite data showing
the temperature difference in
a band around the canal.

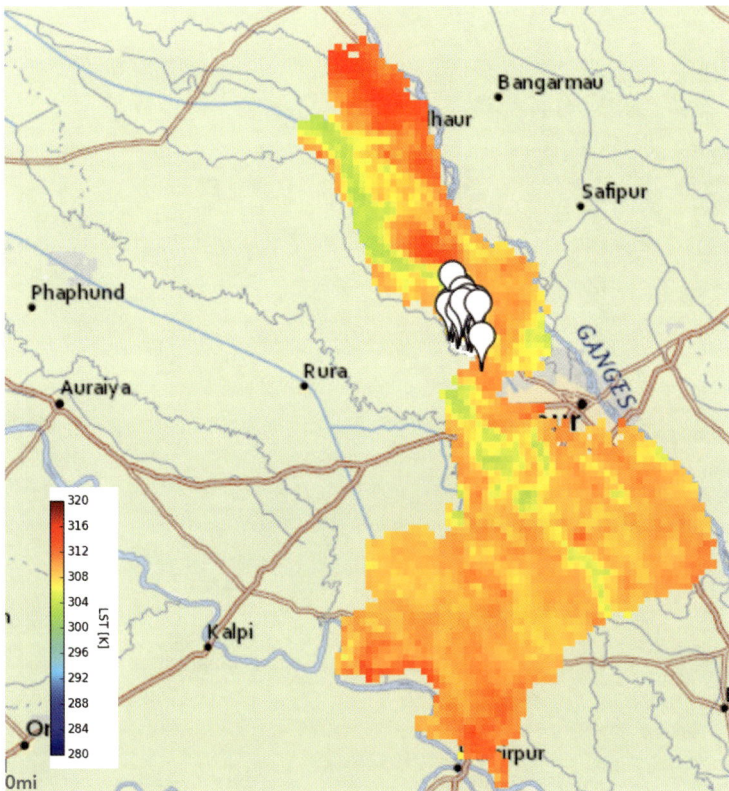

Difference in temperature in October 2017 between north and south
of research area.

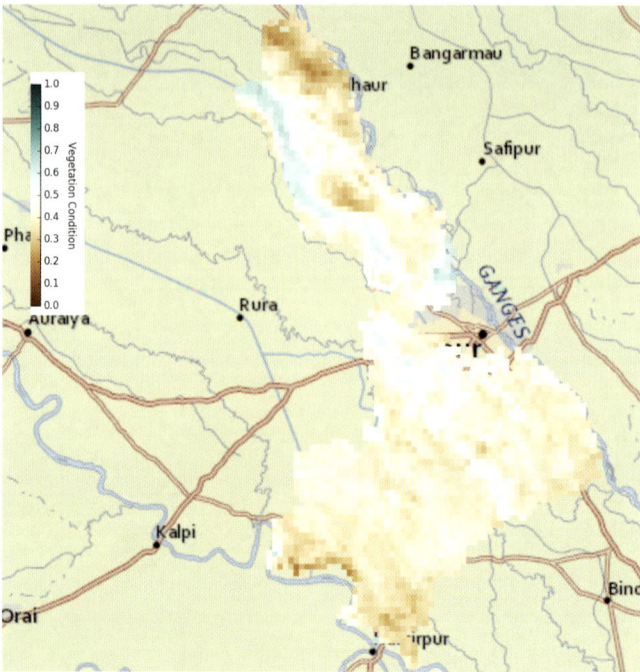

Increase of vegetation health index along the canal.

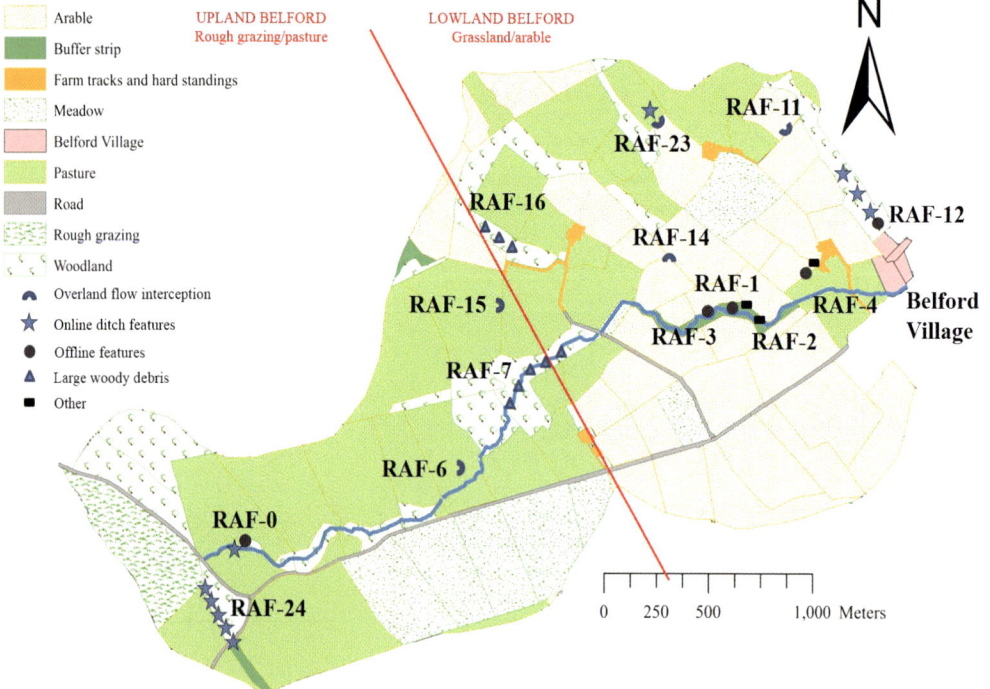

Belford catchment map.

A *pani panchayat* in action. Photo courtesy of Pascale d'Erm.

Jal Sahelis creating a water structure together: Courtesy of Parmarth

Courtesy of Parmarth

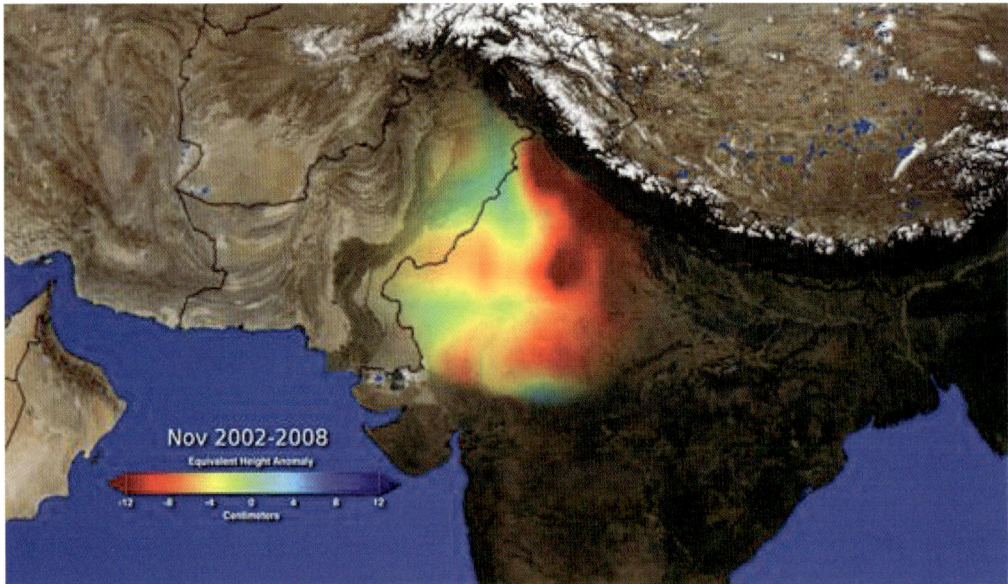

Groundwater depletion of 109 cubic kilometers in three north Indian states, including
Rajasthan. Rodell, M. (2010) NASA Satellites Unlock Secret to Northern India's Vanishing
Water; NASA; https://www.nasa.gov/topics/earth/features/india_water.html

The bank seals the water-holding area. Photo courtesy of The Flow Partnership.

The monsoon rain fills up the structure with 10,000 cubic meters of water.
Photo courtesy of The Flow Partnership.

A lush wheat crop using the water. Photo courtesy of The Flow Partnership.

The stages of the process. Photo courtesy of The Flow Partnership.

a rain garden. The quicker the drain rate, the shallower the raingarden needs to be. A quick-draining soil might need to be 15 centimeters deep and a slow-draining soil 30 centimeters deep. In our case, the chosen depth is 0.3 meter.

Step 1.3: Check the accumulation capacity of the garden according to the volume of rainwater runoff from sealed surfaces. Water can accumulate in huge, potentially destructive flows and needs to be carefully managed. Make sure that the capacity of the rain garden will be sufficient to hold the rain that falls in any single period of prolonged rainfall. If there will be excess water, then it is vital to construct a sturdy overflow channel and a managed way for the excess water to escape. One liter of water weighs one kilogram, so if a significant amount of water collects at one place, then that weight of water needs to be held securely until it can drain into the soil and the aquifer.

Any rainwater project must begin by calculating the likely rainwater volume flowing into the structure and the holding capacity of that structure to determine whether it requires an outlet for heavy rainfall events. In collecting rainwater from a roof, the amount that drains into the garden is the area of the roof. When working in the field, the size of the catchment area is harder to estimate and requires studying the contours of a map to see the area that naturally drains into the site. You can then find out the amount of water that might fall in the most extreme storm event and work out the volume of rain that will naturally drain into the structure.

The overflow structure should be sturdy and should link to a spillway to carry away the excess water. In the case of a rain garden, it is best, if practical, to install a pipe leading to the main drainage system below the rain garden to account for water from excessively heavy storms.

If an exceptional rainfall might reach up to 100 millimeter or 0.1 meter, then 30 square meters of roof x 0.1 meter = 3 cubic meters may be expected to be diverted into the rain garden during such an event.

Therefore each of the rain gardens must have at least a 3-cubic-meter water-holding capacity. Some of this water needs to drain down into the aquifer, and some to evaporate through the plants, so that the holding capacity can empty for a new rain event.

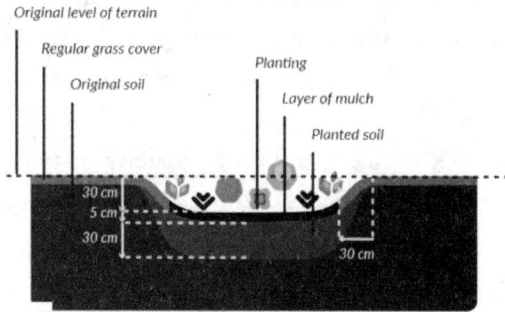

Rain garden design. Courtesy of Peter Bujnak.

Step 2. Design and construction of the rain garden. Digging a rain garden of 3 square meters is not very different from making a *johad* in India that holds 30,000 square meters of water. The sizes may be vastly different but the principles and the problem being solved are the same. Before the construction of the intervention, the soil is compact and dry. When the rain falls, the drainage system from the nearby roof or the geology of a hilly area allows the water to run off before it can penetrate the soil; that rainwater is lost from the local ecosystem. Creating a pond holds the huge amounts of water that fall during a rainstorm or monsoon event, recharging the small water cycle, refilling the aquifer, and irrigating the land.

While in our cities we might not be able to build thousands of huge ponds to revive whole rivers, we can play our part on a small scale by creating numerous rain gardens across the whole town or city and reviving the water cycle there.

Step 3: Situating the rain garden. The rain garden should be situated in a well-drained area or on a very gentle incline (10 percent or less), not too close to the house, and in partial shade if possible.

Mark the area where the rain garden will be located. Image courtesy of Peter Bujnak.

At our school in Slovakia, once the site was chosen, we started by measuring the area of the rain garden. We dug roughly half a meter in and to the required depth. We put the dug soil on a tarpaulin, then cleared it of rocks and debris, turning it into more usable soil for the garden.

An advantage of the smaller, human-scaled rain garden is that we can do the work ourselves, without any machines, and feel how the soil changes when we dig it, becoming receptive to the flowers and shrubs we plant.

Remove the grass surface from the area and dig to the desired depth. Image courtesy of Peter Bujnak.

The ditch should be excavated evenly across the entire area; if the garden is located on a slight slope, the lower side should be slightly elevated to prevent runoff. Image courtesy of Peter Bujnak.

Soil Moisture

The next thing to focus on when making a rain garden is the consistency of the soil. When we started the work, the soil was hard and lifeless. Any rain that fell on such soil was going to simply run off. If you dig a few centimeters down in such soil, you often find it still dry with little, if any, life below the surface. However, when this hard soil is broken up and nutrients are added, the moisture-holding quality of the soil magically returns! The hard surface now becomes porous and the soil can hold the water in the cavities that form in its structure. When it rains, the surface water no longer runs off into puddles elsewhere. The soil soaks up the moisture to support multiple life forms—all of the worms, fungi, and insects that also bind the soil.

This lesson from urban rain gardens is applicable even to huge farming fields. With the use of heavy machinery or cattle grazing late in the year when the field is wet, the soil becomes compacted and the water simply runs off the surface. If the soil is loosened up and full of nutrients, then the nature of the land changes from water-resistant to water-absorbing.

In the case of the rain garden, we can accomplish this by digging out the soil, breaking it down, and adding compost. Healthy soil that is not compacted and has good mineral content—after augmenting with green manures if needed—holds water well. Once the soil has been dug out, all the clumps raked through, and compost added, it can be returned to the rain garden.

Once the water is held in the soil, the moisture and nutrients that it holds become accessible to the plants that draw up the water through their roots. The water is then transformed by photosynthesis and evaporates through the leaves as vapor into the atmosphere. Apart from the pleasure the flowers give when blooming, the vegetation returns the moisture to the water cycle. This has multiple benefits for nature. The soil is enriched, the climate is cooled, and the cycles of nature are invigorated rather than impoverished.

Step 4: Planting the garden. The community then decides what plants are suitable for the available water in the area, planting crops that can survive and thrive so the water table can remain high.

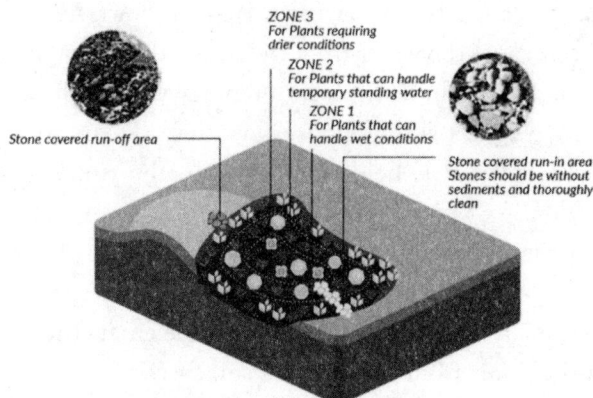

Begin planting vegetation according to the plan. Remember to separate the plants based on their ideal conditions as shown in the diagram. Image courtesy of Peter Bujnak.

Zone 1 wet-tolerant perennials for a temperate climate such as the United Kingdom might include:[7]

- Yellow iris (*Iris pseudocorus*)
- Corkscrew rush (*Juncus effuses*)
- Pendulous sedge (*Carex pendula*)
- Cardinal flower (*Lobelia cardinalis*)
- Arum lily (*Zantedeschia aethiopica*)

Zone 2 and 3 perennials and shrubs tolerant to both wet and dry conditions include herbaceous perennials and shrubs such as:

- Bugle (*Ajuga reptans*)
- Clustered bellflower (*Campanula glomerata*)
- Montbretia (*Crocosmia* "Lucifer")
- Geranium ("Rozanne," "Gerwat")
- Daylily (*Hemerocallis*)
- Siberian Iris (*Iris sibirica*)
- Elder (*Sambucus nigra cultivars*)
- Midwinter Fire (*Cornus sanguinea*)
- Hydrangea ("Annabelle")
- Red Japanese Rose (*Rosa rugose*)

Step 5: Connecting the water inlet. A small downpipe is added to the gutter to transport the water from the roof. When the gutter directs the water through a drain, it will have a tendency to overload runoff in storm events, with the risk of floods. Now that the runoff is led into the garden and the rain is held where it falls, the runoff overload is minimized.

A slope made by stacking up old tires and filling with remaining soil can be used to support the pipe from the roof into the rain garden. Cobbles or gravel around the area where the pipe enters the rain garden help prevent the water from washing the soil away.

Step 6: Maintenance. Properly designed and installed rainwater gardens require minimal standard maintenance. At least once a year, it may be necessary to perform sedimentation removal, erosion prevention, planting of new vegetation, etc. The rest of the year, it can be maintained as one would maintain any normal garden by weeding it and planting seasonally. During periods of drought it might be necessary to irrigate the rainwater garden.

GARDENS BLOOMING IN THE RAIN

In cities, our minds can sometimes separate the plants we see throughout the landscape from the water that is collected in reservoirs and supplied from the tap. Even in a city, a rain garden enables us to practice and experience working with the language of water, reestablishing the link between the rain that falls on a school roof, the moisture in the soil, and the water that is taken up by the plant. This is not just educational—it also helps towns keep floodwaters at bay by slowing the flow of rainfall and storing it during storm events. The features work dynamically to modify the flow of water as it naturally circulates from sky to land to river to soil to trees and back to sky.

The more we engineer our landscape for development, roads, or homes as isolated ends, the greater the need to direct the water that falls in a well-thought-out and deliberate manner. This means shaping the landscape of the town or city to include water as a vital aspect of our shared environment. Rain gardens remind us how pleasing, simple, and necessary such work can be.

If we sit in rooms and try to imagine it, it is hard to see how much effort goes into building even a small rain garden. One advantage of working as a community is that putting in the work together creates a shared sense of ownership. If it is our hard work that helped build it, we feel motivated to look after it forever. We say that everybody who makes decisions on behalf of communities should do some community work for just this reason. When you make decisions together, on the basis of such experience, you are more likely to make the right decisions for the community, for the region, and the planet.

Both the mayor and the regional administrative heads joined in building the rain gardens at the school in Slovakia, and their closing speeches reflected the joy, the pleasure, and the learning the whole process had given them. Both were transformed and acknowledged many times during those two days. Their decision-making for their region was forever changed by this experience.

Building rain gardens so rainwater can infiltrate the ground instead of going down a cement drain to wash away to the sea is an idea much needed globally. A rain garden is like writing a simple sentence using the language of water, expressed through colors, flowers, and patterns. Following simple steps, we can learn to speak this language even with scant access to land or space, bringing all the beauty and cooling effects that rain gardens can offer.

11

Qatar: Regreening the Middle East

From 12,000 to around 7,000 years ago, Qatar was green, with seasonal rains, wild animals, forests, and lakes. Why did the seasonal rains all but dry up? Some scientists say there is a cyclical tilt or "orbital wobble" of the earth that caused the seasonal rains to dry up. Their prediction is that in a few thousand years the rains might return. Another point of view is that it was the human inhabitants of the area who cut the trees, overgrazed the land, burned the fields, and depleted the landscape, effectively breaking the water cycle.

In the absence of dependable seasonal rain, as people located to oases and settlements, a traditional water wisdom around drawing water from the ground developed. Once oil was discovered, running desalination plants meant that the language of water would barely exist for a whole generation, not even in the imagination. But now there is a growing unease in Qatar about their relationship to water in their landscape.

With the global climate shifting unpredictably, the idea that we can control natural processes by the force of oil-run technology is looking increasing shaky. At the very least there needs to be an education program to understand the natural behavior of water.

In conversation with potential colleagues in Qatar, we are told that their grandparents' generation still remembers running through the grasses when they were children, in a world that was still naturally green. The soil had been healthy and full of life. Insects had been plentiful. Beautiful birds would nest in the trees. There were vegetables growing from the carefully managed aquifer water, and wind in the trees. Life had a feeling of abundance then. In their grandparents' generation, growing up rurally, there was a feeling of being connected to it all. The current generation, facing extreme aridity, yearns for that feeling of the country, the sun, the aquifer, the oases, and green places.

Even in the most extreme climatic conditions, it is possible to transform arid areas into oases of lush, green abundance by honoring the relationship between the rain that falls and the land that receives it. We can use this simple yet profound principle across the planet to restore and regenerate ecosystems. The Flow Partnership's partners in the Sahel and Rajasthan, regions at similar latitude to Qatar, are already demonstrating what can be possible in extremely arid conditions.

What secrets do the tradition, culture, and groundwater of Qatar hold for the rejuvenation of their land? Can the ancient systems be revived in the founding of settlements today? How can the mangrove forests that grow in the intertidal zone around the coast contribute to the regreening of Qatar?

TRADITIONAL METHODS OF MANAGING RAINWATER

Until about 70 years ago, the elevated water table of the Arabian Peninsula—a legacy from when the whole Arabian Peninsula was savannah and much wetter—served as a buffer against climate variability.[1]

For hundreds of years, the inhabitants of Qatar lived by accessing large areas of fresh groundwater in the northern and central parts of the country; there was also scattered fresh groundwater in the south.

Towns and early Islamic sites (numbered) throughout Qatar, many of which are now deserted. Courtesy of P. Macumber, 2015, "Water Heritage in Qatar," p. 225. Unesco World Heritage Convention, accessed June 8, 2023, https://www.researchgate.net/publication/304062991_Water_Heritage_in_Qatar.

LINEAR SITES: 1 Al Haddayah; 2. Murwab; 3. Musaykah; 4. Makin 5. Rakayit; 6. Umm al Kilab; 7. Athbar

Numerous settlements developed along the northwestern coasts. Traditionally, settlements were created around geological features called *rawdha* (Arabic for "garden" or "fertile depression"), inland depressions of a hundred meters up to a few kilometers in diameter.

Groundwater under the *rawdha* was accessible through shallow surface wells. The *rawdha* sustained both long-term settlements and a temporary Bedouin presence well into the 20th century. It was only the discovery of oil and the buildup of urbanization that led to the discontinuation of this practice and the abandonment of these settlements.

Rawdha are depressions above geological fractures that take the water down into the aquifer. They are not unlike the *johad* principle of holding water in Rajasthan, India.

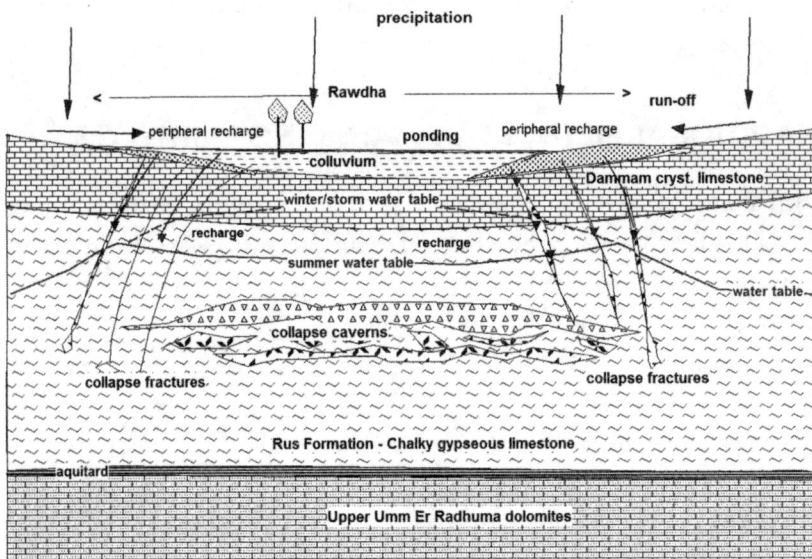

The traditional system of holding the runoff from storms and accessing the groundwater in *rawdha*. Image courtesy of P. Macumber "Water Heritage in Qatar", p.228

Research by Eccelstone and associates[2] shows how the level of the water table in this region is dramatically improved with the presence of a *rawdha* system. When more prolonged rains came, the groundwater levels could rise to the surface. In such circumstances, the *rawdha* is not just a brief ponding effect of the rain but could irrigate an oasis and sustain a settlement.

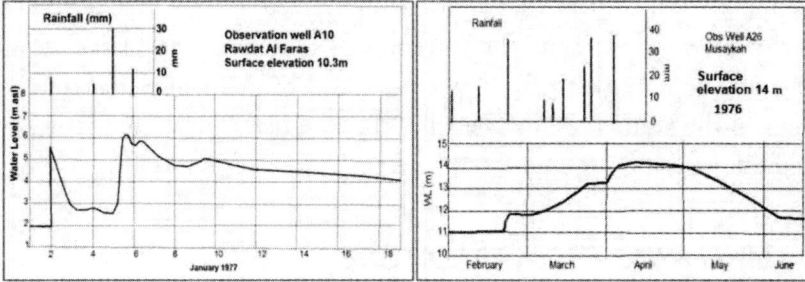

The rise in water table at Rawdat al Faras in response to winter storms (rainfall shown at top). Image courtesy of P. Macumber, "Water Heritage in Qatar", p.229

Restoring the water balance requires reviving these traditional methods. If reinstated, the *rawdha*'s natural depressions would create recharge mounds that would lift the water table high enough to be accessible via shallow hand-dug wells, or even directly accessible from the surface. The villages (*musayki*) could be situated at the edge of the *rawdha*; this layer of fresh water above the brackish and saline water would be a critical resource in their agricultural life.[3]

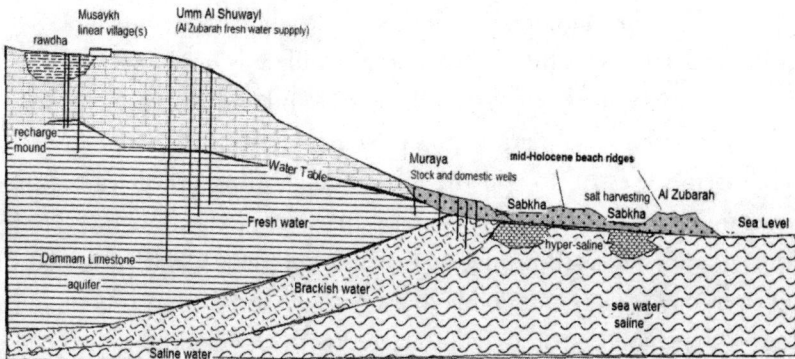

The water situation in Qatar, showing depressions (*rawdha*), villages (*musaykhi*), and salt flats (sabkha). Image courtesy of P. Macumber, "Water Heritage in Qatar," p. 232.

With a long history of traditional rainwater harvesting and careful water management to maintain and replenish these aquifers,[4] Qatar relied on this aquifer water until fairly recently. The above image shows the topological context that allowed surface settlements with access to catchment water. Despite the extremity of the climate in Qatar, a balance was kept between the water used and the replenishment of the water in the aquifers. This seemed to be as much educational as it was methodological.

FALLING AWAY FROM TRADITION

Even 50 or so years ago, there was still a balance between the approximately 50 billion cubic meters of water that people drew from the aquifer and the recharge that the rainwater provided in the same year. But as the groundwater declined, the underground aquifer and wells full of sweet water—the lifeline of the settlements—were overtaken by brackish water incapable of supporting life. Attention to this balance of extraction and replenishment of water—taking just enough to grow needed crops—seems to have vanished.

These days, with the easy availability of treated desalinated water, farmers no longer attend to this balance. Instead, 250 billion cubic meters of water is drawn annually from the aquifer for irrigation—five times more than what the rainfall annually returns. One does not need to be much of a mathematician to see that, with this imbalance, the aquifer is drying out and has reached critically low levels.

Today the water for Qatar's residents and industry comes almost exclusively from desalination plants, which were introduced in 1953. Desalination uses oil to run the energy-demanding processes required to separate salt out of the seawater. Apart from the immense energy requirements and carbon impact of this approach, Qatar's population and economy are left vulnerable to events such as poisonous ocean algal blooms that have occasionally interfered with the desalination process. Desalination has, in a roundabout way, also resulted in a long-term decline in ecological resilience and the capacity for self-sufficiency in the country.[5]

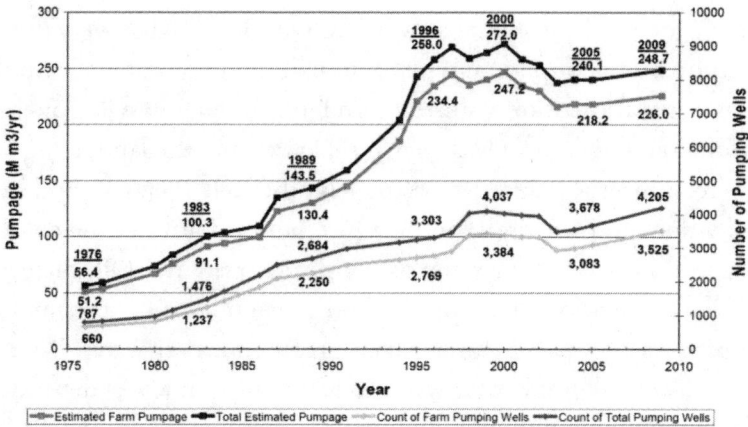

Groundwater abstraction and number of wells in the State of Qatar, 1975–2009. (The line shows the increasing abstraction of water in Qatar, which is now annually around 250 million cubic meters, five times the 50 million cubic meters of water that is recharged.) Image from Y. Mohieldeen, E. Elobaid, and R. Abdalla, 2021, "GIS-based framework for artificial aquifer recharge to secure sustainable strategic water reserves in Qatar arid environment peninsula," *Nature Research Scientific Reports* 11: 18184. Published online September 14, 2021, accessed June 8, 2023, https://www.nature.com/articles/s41598-021-97593-w.

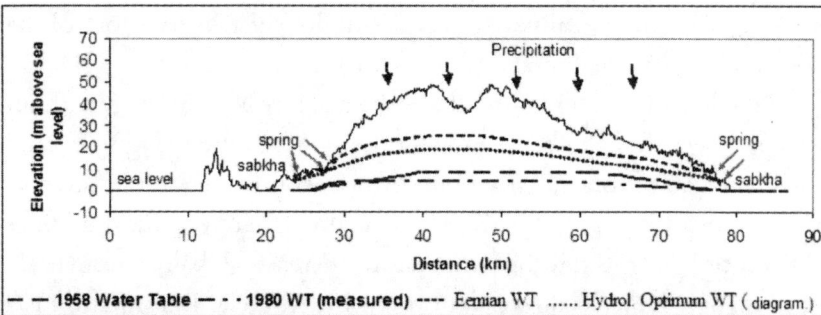

The water table in Northern Qatar. The section through northern Qatar showing actual water table levels in 1958 (steady state) and 1980, and diagrammatic levels during the Eemian and Holocene Hydrological Optimum period.[6,7] Courtesy of Macumber, "A geomorphological and hydrological underpinning for archaeological research in Northern Qatar", 189.

OIL AND WATER

The discovery of oil in 1940 and of the world's largest gas field in 1971 changed the economic outlook of Qatar. The oil and gas deposits are remnants of a time when Qatar was forested. But the discovery did not return the profits of natural reserves to maintain the water system that had once fed those forests. Instead, under the lead of Anglo-American conglomerates, the profit was seen as separate from the land, to be used in overcoming the natural environment through high-rise buildings, desalination, and air-conditioning. Once oil was found in the country, it ruled the day, replacing water as the pivotal resource. Oil, the generational accumulation of organic carbon over millions of years and the dark product of a past life, has taken precedence over water. Whereas one must work skillfully with water to realize life, oil works dependably for its master, who puts it to industrial use. As water enables life, oil spends life. Water requires care, nurturing, and collaboration, while oil can be exploited for advantage, power, and dominion.

Water has become yet another product of the oil-driven world. A desire for instantaneous water at the touch of a lever has replaced the intricacy of dancing harmoniously with the water cycle. This desalinated water no longer tastes of the cool of the aquifer caves, nor does it carry any of the life-enhancing properties of minerals. Something has been cut away from this desalinated water. It no longer tells the story of the land, the country, its life, its culture, or its traditions.

Oil is an addiction that privileges the mastery of technology without a pause for a thought between car rides, air-conditioned offices, and desalinated water on tap. Two billion cars worldwide emit an average of two tons of carbon per year. The machines that keep our buildings cool exacerbate the rate of temperature increase globally through climate change. How far do we want to push that cycle? Desalination is a necessary technical solution for the moment, but the water treatment process is vulnerable to marine algae blooms and is also hugely expensive in terms of electricity.

Investors might see desalination as an easy and limitlessly profitable solution for a limitlessly needed resource. Take the salt out of the water from the vast oceans and seas lapping on our shores and you will have a limitless supply of clean water (albeit without the nutrients that natural water has for our health), a limitless channel of profit for generations to come, and limitless control over the populations of the planet. Water is life and without it no one can survive. Those who control it, control life. How does one go forward from here?

Can the dominance of oil sustain itself through the threshold of the dramatic climate change that we are approaching? The equation does not seem to add up. More and more oil will be required for desalination plants to cool and uphold life as temperatures increase, continuing to fuel the same cycle. How far can we go down this cycle of increasing mastery of nature before we trip ourselves up on this constant demand for luxury, ease, and industrial control? Surely we should also be remembering the place of water, what we have lost, and what we will need for future survival?

MODERN METHODS

In January 2023, unexpected storms produced 100 millimeters of rain near Doha, the capital of Qatar. This is far more than the 40 millimeters of water that fell on Doha in the entire year of 2018.

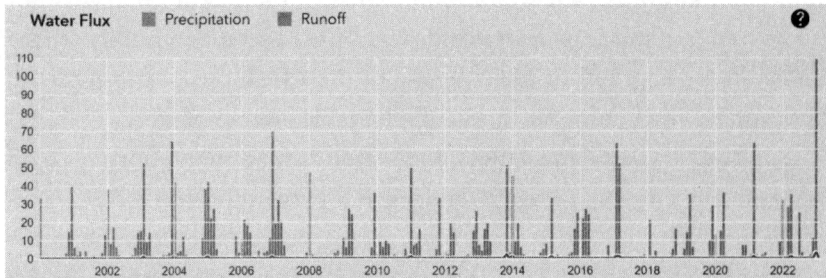

Rainfall January 2023. 109 millimeters of rain was recorded near Doha, showing the unpredictability of climate changes. Image from Water Balance App (GIS), 2023, accessed June 8, 2023, https://livingatlas.arcgis.com/waterbalance/.

In Doha, contrary to the decline of groundwater elsewhere, "the water table beneath the developed city has been dramatically rising since the mid-1970 s as a result of sewage leaks, distillate networks, and septic tank overflow and more recently of losses from the mains carrying potable water as well as excessive irrigation of private gardens."[8] This has led to the absurd circumstance of pure rainwater being pumped out to sea!

Our alienation from water is so profound that instead of having a long-term plan for collecting the water of these January storms to recharge the aquifers and raise the level of water in the wells, a new, expensive engineered construction in Doha has just been completed to transport the water away to the seabed! Mott Macdonald, the engineering company hired for this project, says the following on their website:

> Although Qatar is a desert environment, at least twice a year heavy rainfall results in massive disruption to urban areas and flood damage to infrastructure, retail outlets, and private homes. The vision behind the Musaimeer Pumping Station and Outfall Tunnel in Qatar is to protect the southern Doha area from flooding and future climate risks that could hold back urban development. . . . This has been achieved by pumping rainwater and storm water along a 10.2km-long outfall tunnel and discharging the water offshore on the seabed.[9]

The water cycle is now fully reversed. Qatar drinks the seawater and pumps the fresh rainwater out to sea to be harvested again for desalination! Even knowing that fresh water is scarce, the storm water was not recognized as a precious resource. It was instead seen as a nuisance that needed to be kept out of the concrete of the roads, out of the entrance to shopping malls, and out of the dependable engine of work.

Of course, there must be some water management during flood events in the cities to mitigate damage and prevent daily life from coming to a standstill. At the same time, giving just a little thought to the vast potential of soaking up that disruptive rainwater into the desert would reduce aridity and be extremely beneficial to the climate of Qatar, reviving the respect water had enjoyed in previous generations as a source of life-giving wealth. The website concludes:

> By draining rainwater and alleviating flooding in a 270km² catchment area, the Musaimeer pump and outfall tunnel removes a barrier to sustainable urban development in southern Doha. The new hydraulic infrastructure is designed to keep ground water below –3m, which will reduce the need for expensive foundation works and lower the cost of urbanization.[10]

Desalination is now viewed as the most sustainable solution and seems to have taken over the mindset so much that groundwater is referenced merely as an obstacle to foundation works.

IN THE IMAGINATION

Renewal of a system begins in our imagination—the first place where we open to the landscape becoming water balanced again. Once we take this first step with honesty and integrity, the voices of our grandparents' generation can be heard. If we listen even further back, we find traditions of managing water that maintained the level of the underground aquifer as security for the arid times between rainfalls for centuries.

Several schemes are being investigated now to rectify Qatar's perilous situation, including a scheme by Mohieldeen to artificially recharge the aquifers using desalinated water.[11] This would provide a backup of water reserves in the ground. The technological complexity of these schemes, however, makes one wonder why we are so dismissive of the natural ways that water was held in the aquifers, the vegetation, and the climate through careful tending of the water cycle.

Another challenge is the increasing deterioration of the groundwater quality:

> The factors implying the quality deterioration can be natural such as arid climate conditions, the composition of native rocks and soils, structural pathways for saline water upwelling, seawater intrusion, deposition of salty aerosols, and capillary phenomena, or anthropogenic such as excessive groundwater pumping, frequent irrigation, use of fertilizers, oil and gas production, leakage from wastewater networks, road drainage, and uncontrolled dewatering activities, among others.[12]

This has prompted Kahramaa, the Qatar General Electricity and Water Corporation, to begin building recharge wells and monitoring water use. However, because the knowledge of how to hold the water naturally has been lost, such approaches are small-scale and fragmented. As communities reach despair in deserts such as Qatar, people continue to search for new ways to bring the moisture, the soil, and the trees back into the landscape. Only by making ourselves part of the climate and landscape again will we learn to rebalance our relationship with nature. Even if it seems impossible to imagine that the desert landscape can ever again support life, it is vital as a first step to believe, to communicate, and to discuss the basic principles that can bring the water back into a natural cycle. Even when the way to return the water is unknown, movement comes from imagining it to be possible and renewing the relationship between the community and the landscape. We can no longer abjectly surrender to aridity as the inevitable consequence of climate change on the landscape and rely on technology as the only intermediary. Here the not-so-secret voice of the water cycle can begin to speak again.

With education, people can understand that water is not just a resource existing in isolation to be produced and pumped into taps. Once you have water, you can have trees; once you have trees, the soil can build up; once forests take root, the evapotranspiration cools the surface and produces clouds; once there are clouds, there can be new rainfall. Water is part of a life-giving cycle that must be managed in its different states of transformation, from aquifer resource, to wells, to vegetation, to soil moisture, to clouds, to rain. Nationwide education in the language of water can remind people of a collaboration with the culture of water that has existed for millennia in the land.

We are reminded here of the decade-long, man-made ecological disaster that encompassed 100 million acres in Oklahoma, Texas, Kansas, Colorado, and New Mexico in the 1930s.[13] From 1931 to 1939, around 75 percent of the United States was plagued by unusually high temperatures, the worst drought in a thousand years, strong winds, and resulting clouds of dust. This natural disaster was named the Dust Bowl. Farming practices also contributed to the problem, and things were further complicated by the Great Depression, which took place from 1929 to 1933. By the end of 1934, roughly 35 million acres (14 million hectares) of farmland were ruined, and the topsoil covering 100 million acres (40 million hectares) had blown away. Among other initiatives, the government launched the Civil Conservation Corps (CCC), in which three million young men volunteered for forestry and conservation work. They planted trees, dug ditches, and built reservoirs—work that would contribute to flood control, water conservation, and preventing further soil erosion.

Relief finally came with the rains in 1939, which we say came because of the millions of trees that were planted and the water and soil conservation work that was done by the CCC on that massive footing. What is important in this example is the scale at which a horrific, seemingly irreversible and life-threatening desertification over millions of acres of land was successfully reversed when nature forced political will to bend to its demands. Ken Burns's film *The Dust Bowl* is an eye-popping account of what happened in the 1930s and how the problem was redressed. It is well worth the four hours it takes to view it.

MANGROVE RESTORATION: SIGNS OF HOPE

One example of Qatar beginning to restore the water table can be found in its mangrove forests. Mangroves are marine ecosystem forests growing in seawater. They are ideally suited to climates like Qatar's because they grow in small strips along the coast, having evolved to draw up moisture from the saline margins of the sea. The network of mangrove roots allows the trees to manage tides and create rich wetland environments on its coastal shores, sequestering an abundance of organic carbon.

An important factor in restoring Qatar's intertidal mangrove forests is managing the recharge of groundwater on land, which contributes to the health and greenness of the mangroves.[14]

Although mangroves are notorious for existing in the saline intertidal zone, Matthew Hayes of Villanova University, Pennsylvania, emphasizes that the trees benefit from nonsaline water sources whenever available: "Groundwater flow through coastal wetlands plays an important role in the maintenance of productivity of intertidal ecosystems. Non-saline water sources, such as groundwater and rainfall, are important for the growth and productivity of mangrove forests."[15]

The once ubiquitous presence of mangrove forests along tropical coastlines has been challenged by urbanization and conversion to agricultural land.[16] In Qatar, urbanization has encroached upon the mangrove populations, though thankfully 40 percent of the coastline of Qatar has now been protected and the mangrove forests are being extended.[17] Replanting the mangrove swamps and restoring the groundwater balance to allow the settlement and farming of rural land are ways forward through the harshness of the climate. According to Dr. Mohammad bin Saif Al Kuwari, adviser at the Office of the Minister of Environment and Climate Change, "Mangrove plantations might be grown throughout Qatar's 500-kilometer-long coastline."[18] Like many Middle Eastern countries, Qatar has signed up for an ambitious net zero carbon emission target by 2050, including a plan to plant 10 million trees by 2030.

BLUEPRINT FOR RENEWAL

The history and current state of Qatar is the story of what the world could become. How will we cope if currently temperate parts of the world such as Europe and the United States see rainfall dwindle to 40 millimeters in some years, with violent storms sweeping in and the rainfall running off the hardened surfaces into the sea? Would we rely on desalination or even some future as-yet-unknown technology to master a lifestyle that would let us get by? Or can we look to ancestral wisdom now to reestablish the cycle of nature in the language water speaks, to minimize the possibility of such an uncertain future?

Every country and community across the world has its own wealth of traditions for tending to water in the landscape to support the ecological, social, and economic resilience of local populations. The revival of such traditions and practices can improve ecological resilience, enhance agricultural productivity, and enable far greater water security and self-sufficiency. In The Flow Partnership's experience, sharing between communities across similar climatic conditions can create mutually beneficial partnerships and develop ideas for local water management that can be transposed to new conditions.

The Qatar Foundation, which works in education, research, innovation, and community development, and Caravane Earth, which seeds, practices, and promotes ethical entrepreneurship, are two organizations translating these considerations into applications. A future that returns to a balanced relationship with the groundwater in Qatar is to be found in its cultural tradition. One of the most ubiquitous features found near the *rawdha* sites are the stone remains of simple mosques that used to be dotted around the landscape. These mosques held the story of transformation, both of matter into spirit and of aridity into fertility.

We cannot forget the old cultures. If the challenges of our global society are different today than from a few generations ago, then life demands that we transform our culture to integrate new methods into a whole, viable way of life. How does Qatar draw anew from its experience that spans periods of prosperity of settlements from 12,000 to 7,000 years ago, an increasingly nomadic existence as the climate dried out, and a dependence on technology as the groundwater has depleted?

Qatar's culture grew up through the past millennia of rising and falling civilizations, carrying with it methods to regreen a dry landscape. Both culture and water can be revived as Qatar discovers an era beyond dependence on technology. In Qatar, if rain can be held in the ground where it falls, the groundwater under the *rawdha* can be naturally desalinated. Future generations might know a spring field full of flowers. And we can imagine children playing in those fields, tending expertly to the life-giving water that enables those fields to flourish.

12

Australia: Dreamtime Down Under

To represent Indigenous cultures from most regions of the planet, we are includ-ing a chapter on Australia and the water crisis it faces today. We haven't yet been to Australia to visit a community but have held deep conversations with colleagues and others engaged in water retention work in Australia and have researched independently to gain an understanding of the water situation there. Australia's unique challenges with the financialization of water and water mar-kets are fast becoming a global challenge. The voices of Indigenous traditions and local communities offer strong alternatives to large-scale, top-down solutions for managing climate disaster.

Australia is the hottest continent on the planet and on the front line of climate change. After experiencing the hottest year on record in 2019, Australia was front page news around the world with terrifying film coverage of communities cowering or fleeing from bushfires. The fires epitomized a new era—where climate change was no longer a theory or speculation. With the severity of the Australian bushfires, the world crossed the line into the unimaginable horror of the climate now totally spiraling out of control.

A billion animals were said to have died in the bushfires of 2019, in addition to the almost incalculable number of trees and swathes of veg-etation that burned down, the vast pieces of land that were destroyed, and the loss of human lives, homes, and livelihoods. One of the most heartrending stories that went viral across the world on social media was a video of a dehydrated wild koala bear approaching a group of cyclists, somehow communicating it wanted water from them. They offered the koala some water, of course, and the video has been viewed millions of times across the world since, making viewers achingly aware of the scale of the problem we are now faced with.[1]

In June 2020, a colleague who had been filming projects in India for The Flow Partnership sent us a link to the film *Lords of Water*. The film was especially startling as it stated that the drought and fires in Australia had not prompted research into the underlying causes of water being lost from its ecological system and what urgent fixes could be applied. Instead, the film shone a spotlight on the commodification of water and its trade on the stock markets of the world, so that maximum profit could be made from Australia's dwindling supply. The film takes its title from a statement made by its director, Jérôme Fritel: "The new lords of water live in the city of Melbourne, the business capital. They are bankers, insurers, pension and investment fund managers and they are gradually taking control of this blue gold."[2] In a peculiar twisted sort of reverse logic, the "lords of water" justify this commodification of all life by saying:

"Water falls from the sky, therefore it should be free. Whenever I hear that, I always say diamonds occur in nature and they're not free." —Dean Dray, CEO RBC Capital Markets[3]

"It's a financial product like any other financial product . . . we are just at the beginning of this water financial revolution." —David Williams, banker[4]

"Isn't it a good thing we are finally putting a value on this resource? Because in putting a value on it we're going to respect it more." —Tom Rooney, water market pioneer and CEO, Waterfind[5]

Water is versatile—it changes its form according to the context in which it is carried. Water exists as cloud in the atmosphere, as flow in a river, as moisture in the soil, as raindrops returning to the land. Can't we respect the water cycle simply because of these miraculous transformations, which give us life? Or do we need an external market system of monetization to give water its value?

If water is just a commodity to be controlled, then this is the final irony. The brokers discuss water as if they have injected a new exciting element into the financial system, while playing with our very survival. Water is nothing but a new gamble to be played with, a guessing game of where some water might be left in a land burning dry. If devising a way to make money from farmers by charging them for water is the game we are playing—even at the cost of destroying our ecological cycles—then enterprise will almost certainly be the ultimate winner.

Among all the consequences, perhaps the biggest impact of the new trade will be felt by small farmers who cannot withstand shocks in the water markets. With the water markets gaining traction, farming will be held in the hands of a wealthy few who dictate what crops to grow. Usually crops that use the least amount of water are selected, leading to monocultures that determine what we eat, which isn't always nutritious, diverse, or healthy for anyone or anything but the bank balances of the speculators.

> In our business we want to use our water efficiently because if we do that we can drive a greater profit into the business and that's what we're ultimately here for—is to deliver a profit for the shareholders. —Brendan Barry, water manager at Stahmann Webster, Australia's largest producer of almonds and one of Australia's most well-established agribusiness companies[6]

Fritel puzzles: "Who will come out on top? The planet? The people? Or the markets?"[7] If the markets win in the race to monetize water, then what will be left for us, our children, and grandchildren?

> Yet you all come to us young people for hope. How dare you . . . you have stolen my dreams in my childhood with your empty words and yet I'm one of the lucky ones. People are suffering; people are dying; entire ecosystems are collapsing. We are in the beginning of a mass extinction and all you can talk about is the money and fairy tales of eternal economic growth. How dare you! —Greta Thunberg at the Climate Action Summit, 2019[8]

Over the last six hundred years of colonization and the disruption of so many lands and their people, the approach bringing us to this point has not changed. What was it like in the 1700s and 1800s when native Aboriginal people were still around on the land before the settlers arrived in Australia?

INDIGENOUS WISDOM

In Australian Aboriginal culture, it was clearly understood that water was sacred. Aboriginal religion and culture have their foundation in Dreamtime, dating back to over 65,000 years ago. The Aboriginals believed that everything present today was a creation of the ancestors— the rocks, the streams, the rivers, the animals, and the whole world were made by the ancestors and given to us by the spirits. These spirits and ancestors watched over us from sacred places. In order to live good lives and make them happy, the Aboriginal people were tasked with looking after these gifts. Dreamtime is thought of as a period on a continuum of the past, present, and future—without beginning and without end.[9]

The Indigenous Knowledge Institute of the University of Melbourne states:

> Although it may seem commonplace to distinguish between land and water, in many Aboriginal and Torres Strait Islander cultures, land and water constitute a single cultural landscape. This landscape links the sky, the rain that falls down to flow into the rivers, and the water that seeps into the ground. This water (recharge) travels slowly (sometimes taking millions of years) through porous rocks, and then reappears (discharge) as flows to rivers, natural springs, shallow aquifers, soaks, or mound springs. The diversity of environmental regions and precipitation across the continent and islands was matched by Indigenous ingenuity in adapting their social and technological systems to the variability of water availability.
>
> Aboriginal and Torres Strait Islander people established a complex knowledge base that enabled their survival on one of the driest continents. This in-depth understanding of the hydrology and hydrogeology in each country, and also regionally, allowed people to find and re-find water in the landscape. This may have

involved finding deep pools or billabongs when the rivers were not flowing, or protecting rock holes from animals by creating purpose-built lids. Rules and laws were made to protect the quality of water, particularly in cases where the water was for human consumption. For example, one way to protect the potable water is to only drink from the bottom pool, not the top pool, to avoid contaminating both water sources.[10]

Different Aboriginal stories of water and its sacredness from Dreamtime are represented by sweet creatures, sometimes real and sometimes mythical, holding the mantle of water—Tiddalik (the water-holding frog in the Dreamtime creation story), the bunyip (a mythological swamp monster), the Ngapa Jukurrpa (ancestral water dreaming rainmakers), Baiame (the father supreme spirit who created the rivers, sky forests, songs, and traditions), and so on. Their stories contain disaster, life, death, humor, kindness, and a deep respect for that which gives life: water. The stories also teach, subtly and not so subtly, what happens when responsibility for protecting that life-giving source is disrespected or ignored.

One book that covers some of the history of Australia that isn't typically taught in schools—and which has been met with much controversy—is Bruce Pascoe's *Dark Emu*, which argues that the Australian nation should embrace the sophistication of the Aboriginal people in their culture and lives as opposed to labeling them as uncultured, ignorant hunter gatherers.[11]

Australian reporter Rafqa Touma, who reviewed *Dark Emu* in *The Guardian*, writes:

Pascoe's book reframed the colonial lens through which Australia understood its Indigenous history, which had reduced Aboriginal people to simple hunter-gatherers prior to European settlement. Pascoe set out to prove ancient Aboriginal practices of agriculture and construction, which showed a sophisticated use of land that had predated colonizers. Framing Aboriginal people as hunter-gatherers implied they only had a loose connection to land—a convenient position for those who wished to colonize it.

Anthropologist Marcia Langton credits the book for having shifted the perception [in Australia] that Indigenous people were "savage" before colonization and became "fully fledged human beings" after.[12]

In the book, Pascoe uses settler writings from the late 1700s and early 1800s to shine light on Aboriginal people's profound understanding of the well-being of the land and how to tend to it. He writes that "the Aboriginal methods of land management were not just practical but aesthetically pleasing. At the time, the surveyor general of New South Wales, Major Thomas Mitchell, noticed the beauty of the country but considered it an accident."[13]

We crossed a beautiful plain; covered with shining verdure, and ornamented with trees, which, although "dropt in nature's careless haste," gave the country the appearance of an extensive park.
—Major Thomas Mitchell (1792–1855)[14]

The Indigenous relationship with the land surprised even modern Australian historians like Rupert Gerritsen (1953–2013), who wrote: "More had been done to secure provision from the ground by hard manual labor than I could believe it in the power of uncivilized man to accomplish."[15]

When one thinks of modern rural Australia one imagines sheep by the thousands, perhaps not understanding that it was the colonists who brought sheep to the continent. In 1788, the first 29 or so sheep arrived in New South Wales, Australia, on the first fleet of ships with convicts—now called settlers—from the United Kingdom. Charles Darwin, whose ship the Beagle was harboured in Sydney Cove from January to March 1836, wrote of the savannah bushland encountered beyond the Blue Mountains:

The woodland is generally so open that a person on horseback can gallop through it; it is traversed by a few flat bottomed valleys, which are green & free from trees; in such spots the scenery was like that of a park & pretty.[16]

John Rickard refers to a number of colonist authors referring to the land as "lush and green," ignorant of the fact that this "park like appearance" had been purposefully cultivated by Aboriginal people with the systematic use of fire.[17]

In the 1800s, sheep farming began in earnest to create a wool and meat industry. Today there are between 65 to 75 million sheep in Australia. Sheep have unique teeth that enable them to bite the grasses down to the point where the grasses cannot regrow. Over time, the native grasses died out, leaving vast drylands. In these Aboriginal lands occupied by settlers, sheep farms began planting a new kind of engineered grass that was tough and resilient and could be artificially irrigated. Pascoe, referencing *A Million Wild Acres* by Eric Rolls, documents these disastrous effects:

> In his epic, *A Million Wild Acres*, Eric Rolls described the desecration by sheep of the grasslands in the Hunter-Pillaga region. Rolls was a passionate man of the land who documented the misuse of soils and water by Australian farmers. He noticed that the dispossession of Aboriginal people and destruction of their villages was followed by an equally rapid deterioration in the soil, the foundation of the pre-contact economy.[18] Rolls writes, "In Australia thousands of years of grass and soil changed in a few years. The spongy soil grew hard, the runoff accelerated, and different grasses dominated."[19]

The hoeing of the land by the Aboriginal people was replaced by mechanized farming by the colonists. Although this was a great leap forward in making profit from the land, Pascoe argues that the practice came at a cost. Over 65,000 years of tradition in managing the land was unwittingly overwritten with an imported colonial model of farming:

> The fertility encouraged by the careful husbandry of the soil was destroyed in just a few seasons. The English pastoralists weren't to know that the fertility they extolled on first entering the country was the result of careful management, and cultural myopia ensured that even as the nature of the country changed, they would never

blame their own form of agriculture for that devastation.[20]

Australian settlers saw themselves as owners of the prosperity they brought and hence entitled to profit when taking over the land of the Aboriginal people. One Indigenous Nagarundjeri elder called Uncle Muggie has this to say:

> Water for me and my people is a part of Who We Are. It's a part of our stories, our creation stories. Selling? Trading? What is it? Filling your dams up, making sure that you've got your share, making sure that no one else takes yours . . . part of it is madness and most of it is greed. Greedy people—they want the water but they also want [all] that [they can grab]. Then they will sell it to somebody else for money. You can't eat money, you can't drink money.[21]

Turning a profit from what gives life at the cost of losing life is a possible end result of the wager that is being made. It is up to us to understand the incalculable value of our planet and its climate, of holding water in the land by whatever means possible. It is proven over and over and over again that trees and water influence and regulate the climate and cool the planet. Communities who speak that language of water keep our planet healthy.

Dark Emu and the stories of the explorers reveal how colonization could only justify itself by seeing its intervention as wholly required. The new settlers insisted that culture in this place "began" with colonization. References to prior cultures and ways of relating to the land were erased from the history books.

As the land dries up and rain no longer falls, or the fires consume the forests, this unwavering faith in modern technology as being the "solution" has led to an "answer" that commodifies even water as a tradable stock! What kind of thinking is that?

THE DECLINE OF THE COMMONS

The system of privatizing every aspect of the water structure comes to us from Margaret Thatcher, prime minister of the United Kingdom from

1979 to 1990. Her thinking was that everything including the basics of life could be traded. It was a view that came to dominate economic policy in many countries in the next few decades, opening the flood-gates of profiting from water and controlling it for maximum gain. And nowhere has it been taken on as enthusiastically as in Australia, where financial markets have started a revolution to make water profitable and create water markets just like oil markets.

> Water scarcity is part of the future of the world. The global pre-dictions are that by 2050, more than half the world will be living with limited water resources and abundance is a thing of the past.
> —Mike Young, University of Adelaide, Founding Father of the Australia Water Markets[22]

Is this the final frontier? When we can no longer depend on water from the sky as being a gift from nature and our very act of staying alive depends on some hidden trader whose finger is on the click of the app that sells us our water? What of the people who won't have money to buy this "commodity"? How far will humanity go chasing zeroes indicat-ing profit on a screen, and how long will those zeroes hydrate us and all we hold dear in the name of our children and grandchildren?

Professor David Hall from the University of Greenwich has kept his eye on what is being done in the name of privatization of water since the early 2000s:

> What happened was the entire system, the entire physical system as well as the concession was sold to the private companies. One of the first things some of the companies did was to start cut-ting off water supply to people who hadn't paid their bills. One company for example disconnected 11,000 customers and as far as the company was concerned, they could stay disconnected: if they didn't pay their bills, they didn't get any water. From the market perspective that doesn't matter. The market can't care less if people die from cholera. Really that's not their job. Their job is making money and they've done that very well.[23]

It took years before the first law prohibiting companies from turning off someone's water was passed.

As a continent, Australia is never far from drought, and the powers that be and scientists are all looking at water security and how to counteract these droughts and the threat of running out of water. In such a situation, "desalination" is a buzzword that's never far away, and Australia is no exception. In the logic of technological progression, the defects in the current situation can be solved by future inventions as yet unknown to us. For instance, the lack of planetary water as a "commodity" will be solved by a new era of desalination plants all around the coast. However, as climate change is making apparent, this kicking the can down the road kind of thinking will never succeed in outwitting nature. The intervention to hold the water has to be built in our own backyards today if we want to grow food from that water tomorrow.

There is a peculiar disconnect among those who chase zeroes and play the market. These vague, shadowy figures that seek to be the Lords of Life are human beings too; they haven't been doing this consciously with evil intent. Yet somehow they cannot connect their actions in the chase for profit to the wind that blows through their hair when windsurfing or sailing or walking across a lush mountain.

It starts with wanting a little more than what you already have, which soon gathers momentum. When those who have what you want will not give it up willingly, you devise any means necessary to wrest it from them. You can see how all that spirals out of control with the effects of colonialism throughout the world. There is no one particular figure driving this disconnect with a plan to destroy the planet. It all comes about with an interlinked series of actions and events extending back to our forebears and compounding with every generation. In the same way, regeneration and making the planet healthy again will come with our collective, interlinked actions—which can begin right now! Water is a language that we all have the capacity to speak, even if many of us don't speak much of it now. It's like exercising a muscle—if you don't exercise it, then we begin to lose it. We need to exercise that muscle again or we will lose it forever.

THE DEARTH OF INDIGENOUS CARE

As we began to investigate what was happening on the Australian continent, the first question we posed was "Are there any Aboriginal-owned
and -managed water communities in Australia today? If so, who are they
and can we contact them?" The answer was swift—none. Water in Australia is managed by the government.[24] There are few community projects looking into water management and none that we could find were
owned and managed by those native to the land. In the process of edit-
ing this book, we came across the work done by Anna Poelina with the
Martuwarra Fitzroy River Council on the Indigenous cultural approach
to collaborative water governance.[25]

A colleague, in a private email, described this:

> The Aussie legal/regulatory regime is unsupportive of Aboriginal
> activities that are "off the grid" (not recognized by or administered
> in conjunction with State authorities). Those Indigenous efforts
> at managing water that do happen are otherwise vulnerable to
> attack or are being undermined. That may account for why exam-
> ples are hard to find, because the impacted communities want to
> keep a low profile.

The Indigenous Knowledge Institute of the University of Melbourne explains:

> Since colonization, Aboriginal peoples have not typically been
> included in the management and distribution of water. This is
> mostly because other parties such as industries and irrigators have
> more time and resources to devote to understanding complex
> water legislation and attending meetings to assert their needs.
> Due to a history of dispossession and marginalization, relatively
> few Aboriginal people are landholders, with fewer resources to be
> vocal about their water needs and a diminished sense of power in
> the community. In addition, historically, arguments in favor of
> sustainability over profit have been difficult to sell.[26]

After much searching, we found one instance today where the Aboriginal tribes of the area are being consulted and are participating in decision making, even though the entire project is not in local community hands. The Coborg peninsula, an area of 2,100 square kilometers, with only 20 to 30 people living in it, is a part of the Garig Gunak Barlu National Park. The Australian government's Department of the Environment says:

> The first ever Ramsar-listed site (The Convention on Wetlands)[27], Cobourg Peninsula in the Northern Territory, has unique biodiversity, landforms, habitats, and wildlife including terrestrial, riverine, freshwater, brackish, and coastal/marine ecosystems. Indigenous people have lived on the Peninsula for over 40,000 years. It is considered that the Creation Ancestors first entered Australia via Malay Bay near the Cobourg Peninsula before travelling across the rest of the country creating people and places. The on-going role of the Traditional Owners (the Arrarrkbi) in the joint management of the site has helped to maintain its natural and cultural values.[28]

It is ironic that the Aboriginal care of the land is honored as a tourist attraction today without regard for the underlying traditions that were able to respect and care for nature and its gifts to life.

The Australian government manages the water and delivers water infrastructure projects in collaboration with communities with a top-down approach, which is useful up to a point. One could argue that with the situation becoming so dire and desertification becoming the most pressing issue for the continent, it is vital that there be a measure of control by the government to enable some kind of big-picture thinking and regeneration.

With the above in mind, the case of the regeneration of the Ord river is noteworthy.[29] By rediscovering Indigenous methods of looking after the landscape by holding water and strategic fencing of lands as key to keeping Rickard's landscape "lush and green," sheep- and cattle-grazed degraded lands dating from settlements in the 1880s in East Kimberly were regenerated.

Despite these individual instances of regenerating land degraded by overgrazing, the basic Indigenous ways of caring for the water, soil, and land seem to be sparsely followed.

Of course, there are also land practices that are being tried successfully by individual Indigenous farmers and we look forward to continuing to learn more.

STRENGTHENING THE MUSCLE: NATURAL SEQUENCE FARMING AND PETER ANDREWS

Peter Andrews, who would become the founder of Natural Sequence Farming (NSF), a rural landscape management technique aimed at restoring natural water cycles that allow the land to flourish despite drought conditions, spent 40 years nursing a property called Tarwyn Park. According to the Tarwyn Park website, the park was a "run-down, salinized, seriously eroded, and degraded former horse stud farm. Like most Australian farming land, Tarwyn Park had been faced with over two hundred and thirty years of clearing, burning, farming, overgrazing, and draining of the landscape by our forebears, which as a result, had produced a very different landscape to the one that was present in 1788."[30] Using the principles of NSF, Andrews was able to maintain green and fertile pastures while neighboring properties dried out in the midst of drought.[31] His documented results of reduced erosion, adequate water driving plant growth, and the return of native biodiversity are impressive.[32]

Andrews' NSF method offers a low-cost, widely applicable method for reducing drought severity and boosting productivity on Australia's farms and landscapes. The technique is based on ecological principles, low input requirements, and natural cycling of water and nutrients to make the land more resilient. This is what the practice of restoring degraded Australian landscapes to how they would have been prior to European settlement can look like.

A FUNDAMENTAL RIGHT

With rising alarm about the increasing financialization of water, Maude Barlow, a Canadian human rights activist, lobbied for and won a most crucial victory for the planet when the General Assembly of the United Nations in 2010 declared access to water to be a fundamental human right, a vote later ratified by all member countries. In her speech, she focused on the urgency of this question:

> There are two paths and I don't know who is going to win this. It is either water as a commodity and put on the open markets or water is understood as a human right. You can't have it both ways. You have to choose now. —Jérôme Fritel, *Lords of Water*[33]

> The United Nations calls water scarcity the scourge of the Earth. To truly guarantee the human right to water we must protect it as a public trust and as a commons, not a commodity to be put on the open market for sale like oil and gas. And we must challenge the current power structures and institutions that support unequal access to the planet's dwindling water supplies. Our goal must be clean, affordable, accessible public water for all, for everywhere, for all time. —Maude Barlow, Stockholm 2018 Conference on the Future of Water[34]

On July 28, 2010, through Resolution 64/292, the UN General Assembly explicitly chose to recognize the human right to water and sanitation and acknowledged that clean drinking water and sanitation are essential to the realization of all human rights.[35]

Maude Barlow, we salute your prescience and visionary actions in fighting for water as a fundamental right!

COLLECTIVE ACTION

We are living on the Dreamtime continuum of past, present, and future. Australia offers us a glimpse and a warning of what is to come if we don't start speaking the language of water. Water poverty is not an "us versus them" divide—we are all in a state of water poverty as the ecosystems of the planet are breaking down. And the multiple communities we describe in this book, who are holding water and changing their lives and regions locally, are giving us the solutions to change the path we are on at a planetary scale.

Dreamtime does not mean returning to an Indigenous lifestyle. The essence of it is allowing humanity the standing to access its own future in the unity and harmony of existence, rather than in the fragmentation of greed.

13

Bundelkhand, India: Making Friends with Water

This chapter is based on visits, meetings, and conversations with three Indian women—Sri Kunwar, Phoolvati, and Neelam Jha—who are part of a revolutionary movement of Indian village women who are restoring their local water sources and managing them efficiently. The movement is called Jal Sahelis, which means "women friends of water." The women wear bright blue saris to signify the raindrop, which is what they have become for their villages. This chapter is to be read keeping in mind that often gender-specific roles for men and women exist in traditional societies across the world. In this chapter, we look at how the women overturned 4,000 years of such historical limitation to bring greater equality between men and women in an extremely conservative society. The whole dynamic of the community changed for the better as a result of speaking the language of water. We've deliberately kept the language simple in a direct translation of the women's thoughts, expressed with honesty and charm.

Is the language of water feminine or masculine? In India (and in other parts of the world too) local tradition declares that fetching water is women's work. The woman is the one who creates the home, who gives birth to the children, gives them baths, sends them to school, prepares the food for the family, and looks after all of them. And she needs water for that, no matter how far she has to travel to get that water; otherwise she cannot make a home or do her work. The same tradition says that digging ponds, laboring in the landscape, deciding infrastructure requirements, and making the decisions is men's work.

Rural societies (especially those in the villages of middle India) frown on mixing gender roles. After all, there is the weight of centuries of tradition behind all these customs that keep the boundaries drawn between men and women. In this chapter we will go beyond these social and cultural norms, to tease out the qualities of water in bringing about deep social change along with changes in the landscape.

THE WOMEN

It started with a woman called Sri Kunwar in the village of Udguwan.[1] Four feet, 10 inches in height, very slim and slight as if a puff of wind could blow her away, fiercely determined, she is a doer. Her slight build belies a will and determination that carved a pioneering path that brought water to her village and helped the Jal Sahelis (Friends of Water), a women's organization working on water issues in these villages, to come into being.

Sri Kunwar got married at the age of seven, knowing nothing about what marriage meant. Child marriage was banned then, but who followed that in the rural areas? She was married off to a young man in a poor family where there was often no food to eat. Circumstances had been somewhat similar in her parents' home. Life in her married home was harder, though. She wasn't allowed to go out and play anymore and had to help her mother-in-law and other elders to run the home. She spent a large part of the day going down to the well, about a mile away, and getting water for the household chores. She had to carry the water in pots on her head or in her arms. There was no canal running nearby, no hand pumps, no pipes, no taps bringing the water home. This was attributed to the disappearing rains. She had to go and fetch whatever little water she could from that well far away from the house. Here was a seven-year-old girl, condemned to a life of drudgery before she even had a chance to grow up.

Yet even at that young age, and of course in the ensuing years as she grew into a young woman, Sri Kunwar said she knew that giving life and creating a home was women's work. She had to go and fetch the water even when she was pregnant; sadly, she lost her child from carrying too much weight without enough strength in her body.

During all those years of fetching water over many miles, a determination grew to improve her lot. Surely there was more to her life than this seemingly never-ending trudging with pots of water on her tender head? Every year, as she watched the rain fall and flow away, she wondered why they weren't keeping that water. This led to wondering how she could keep that water for herself so she did not have to carry it on her head for miles, and how she could use that water in the fields. If no one else was going to stop the water from flowing away, she would have to do it herself. If she stopped that rainwater, then it would go into the ground and the wells. While tramping the fields back and forth with a heavy pot of water on her head, she had come to this simple understanding of hydrology.

She asked the men in the village how the rainwater could be stopped. They all spoke vaguely about check dams, dismissing her, saying that it was not for her to think these thoughts and neither was it women's work. Once she even went inside the well to learn more about how it was built and perhaps build one closer to her home. She became more and more determined to build a water-holding pond for herself and her village so she could have access to water when and where she wanted it. How was she going to do that if no one was supporting her? She said that she decided to do whatever it took, even if it she had to become the laborer creating the pond.

Hearing of her crazy ideas, some of her village's women told her about the NGO Parmarth Social Service Organization (PSSS)[2], which was solving water issues with the people of that area. They encouraged her to talk with the people at PSSS to learn more about how to act on her determination. If she was crazy enough to think she could build a pond or check dam by herself, then she must at least talk with PSSS and go about it the right way.

Phoolvati was another strong young woman, living in another village nearby called Hanauta.[3] She was younger than Sri Kunwar and even poorer, living an even harder life. Phoolvati said that she couldn't study much due to their poverty and the lack of water at home, even though she really wanted to go to school. Her parents just could not afford to send her to school. Phoolvati got married at the age of 12. What does a 12-year-old child know? Phoolvati had no idea how to be in a formal relationship with her in-laws or what it meant to be a daughter-in-law, let alone have a deeper awareness of water sources and their meaning for her life.

Phoolvati faced similar water problems at her husband's house that she had faced at her father's house while growing up. There was no metal vessel to store the water, just one mud pot, which Phoolvati took to the well. During the day, if the water in the pot was finished and somebody needed water, she had to drop whatever she was doing and go to the well again to fill it up and bring it back. The well was very far and the pot was very heavy and Phoolvati had to fetch water many times every day. As a girl she had no voice in any aspect of how her life and time was to be lived. All she had every day was unceasing work, entirely at the beck and call of the men and elders in the family. All her dreams of education were completely dashed, vanishing along with the rain in her region.

As she grew older, she realized even more keenly that women not only had no voice, they were not even credited with any intelligence or sense. They were not allowed to say anything, talk to anyone about the water or any other problems they were facing, or be part of any decision-making in resolving these problems. They weren't allowed to go anywhere—just to the well and back.

On one such trip to the well, when Phoolvati heard about a meeting that was being organized by PSSS, she plucked up her courage and asked what the meeting was about. When she was told that the meeting was about water, Phoolvati was determined to attend it. She asked for permission from her mother-in-law, who surprisingly said yes, but her husband refused to let her go, saying that she needed to work in the fields and not go gadding about attending useless meetings (and anyway, such meetings would give her ideas above her station). Moreover, there would be other men at the meetings and it was a social taboo to be in the same place as other men unaccompanied by male members of her own family. But Phoolvati fought with him and went to the meeting anyway. When she came back, he beat her for disobeying him, but she says it was worth it! Phoolvati told him that he did not have to carry the water so he had no clue how hard it was. All he knew how to do was to have a bath using the water that Phoolvati fetched on her head multiple times a day, eat the food that Phoolvati cooked with that water, drink his tea made with the water that Phoolvati carried from the well, and leave for work. He beat Phoolvati even more for saying all that and, in his view, being insolent with him.

In a third village, called Rund Ballora[4], lived Neelam Jha, whose life was unique in that she was born in a small town and had gone to school before landing in a rural village as a bride. She says:

> When I was five years old, I saw that towns were good places to live in. There was water and other facilities. When I got married at the age of 16 and came to this village, I had to wear the veil and go long distances to fetch water. I was married off at the age of 16 because my mother died and my father wanted me to get settled in my own home quickly.
>
> Very soon after my mother's death, one day people came to my house. My family did not tell me that these people had come to see if I would be a suitable bride for their son. There were a lot of people in the house, and I was asked by my family elders to make and serve tea to all of them. The visitors liked me and, only then, my father and brother told me I was to get married within the week! My father was keen that I get married to this particular man as he was an only child and hence considered a good match. I had to leave my studies halfway to get married, which I did not want to do at all but I had no say in the matter.

Neelam's husband was 26 years old, 10 years older than her but with a similar level of education. Her father- and mother-in-law were uneducated. Her new life was a world apart from her last:

> Once I got married and came to my husband's house, I realized it was in a very rural and quiet area and completely different from where I had lived before getting married. There were no vehicles or transport, no roads, and no social or cultural life like there is in towns and cities.

I could only go out to fetch water or to use the toilet, which was also new to me as we had a toilet in my father's house. There was no freedom at all. All my freedom had been taken away. I was made to cook on an open fire, which I had never done before, having had piped natural gas in the kitchen in my town home. I really wanted to complete my studies and become a graduate and thankfully my mother-in-law knew that I was facing adjustment troubles to this new married life and supported me in that. She quietly saved money from the household budget and helped me complete my studies. I was lucky to get a mother-in-law like her!

To fetch water I had to walk 1.5 kilometers to the hand pump each time. I would have to carry two or three pots full of water on my head, one on top of the other. Each pot held about 10 to 15 liters of water. Each time I was carrying approximately 35 liters of water, six to seven times a day! This was a new thing for me. In my life I had never had the task of fetching water like this. In addition, I had to wear a sari (when I was used to wearing Indian summer pants instead) and cover my face with the veil. And the veil! I had never worn a veil and I had to wear a veil that was drawn over my face so low that it covered my whole face and I could barely see when I walked! That was the custom for married women in the village. Because of this veil, I could barely see when I walked with such a heavy load. It was so difficult that I fell down many times.

At home every day there was a fight as I could not finish the cooking on time and everything would get delayed. I had so much work to do and going again and again to fetch water, to wash clothes, bathe, water the farm—it was an impossible, unbearable, and untenable situation for me. I almost felt like giving up on life.

And then I became pregnant. Even during my pregnancy I had to work on the land, look after the house, the cows, my studies, and bring the water. There was no respite from all of that. My first child was a daughter. Even at that time and not being so strong post-delivery of the baby, I had to continue working in that same way as before, with all the responsibility still on me. I really felt ill during that time. I had a stomachache and the doctor said I should not pick up any weight but I wasn't allowed to rest as there was no one else who would do the work in the house. Somehow I had to find the strength to carry on and make peace with all this hardship. I was educated and yet I was being treated like an uneducated person who had no intelligence and was just a beast of burden. I felt I had gone backwards in my life. This was the most difficult period in my life.

Then in 2008 my husband became quite ill. That same year, my younger daughter was also born. Financially our situation became really bad. During that time, I kept thinking I should get a job outside and earn some money. After all, I was educated up to a point. I was definitely capable, but what could I do in a village? Was there any work that I could do within the social norms of the village society? I couldn't work as a laborer, having had two operations when having my children. My body was not strong enough for that kind of hard manual labor. Nor did I want to do that kind of work.

Then one day I saw a group of women sitting and talking and I thought that I should go and listen to them and ask them how they were dealing with their problems. Were they having similar problems to mine? Talking to them eased my mind. It was a ray of light for me and after that I joined their women's self-help group and was often able to be part of the meetings with them. In those meetings we would talk with each other and share our problems and pain. I made friends. I also spoke about the water problems in the village, but they all said that there was no money to solve the water problems. I suggested we talk to the village headman about doing something. When we spoke with him, he said there were no schemes yet for doing any work on the water situation and if any schemes came he would tell us. But of course nothing came from him. We were also dismissed as "chattering women" and not taken seriously.

Some time later the local coordinator asked the women's group for a person to work on a UNICEF project for reducing mother and child mortality. I was selected for the job since I was the only one who was educated. I was responsible for 25 gram panchayats[5] to train women on how to deal with their problems at home. This was my first job and I was very happy. But of course it did not solve my problems of the daily grind and trudge to get water every day—just added a slightly more fulfilling dimension to my life.

I heard from somewhere about PSSS and its head, a man called Mr. Sanjay Singh, and the work he and his team were doing with communities on water retention in the villages. In the next meeting with the women's group, I suggested that we find out more about PSSS. Maybe a solution to our water problems might come from them.

I had been working with the UNICEF project for 1.5 years before I met Mr. Sanjay Singh. I was a bit scared to speak with him at first since he seemed very accomplished, but the women put me forward to speak with him. He was very kind, spoke in detail with me, and then arranged for me to start a water solutions committee in my village and so I lost all my fear of him! I set up the women's Pani Panchayat Samiti (water solutions committee) in my village (Durgapur) in 2018. In this committee, the main problems we discussed were:

- Our children couldn't go to school on time as we were busy fetching water.
- Our husbands couldn't go to work on time as we hadn't been able to cook their meals on time.
- Lots of strife and fights in the family because of these water-related problems.
- Women got scolded and beaten for minor things that were not their fault.
- Daughters were unable to go to school. The mothers were determined that their daughters should not face the same problems they had and should get an education.
- Women suffering similar hardships.
- Pregnant women falling when fetching water and often becoming ill or, in some cases, dying from accidents due to carrying very heavy pots. That was a horrible and sad thing.
- We needed water in the home, rather than fetching it from outside.

How could we resolve these problems? How could we bring water into our homes? Every day I wished that our next generation didn't have to face this water scarcity that left us women emotionally and physically drained. Is it a crime to be a woman?

THE VESSEL APPEARS

Sanjay Singh, the leader of Parmarth Samaj Sevi Sansthan (PSSS), is a fairly young man and an enthusiastic and visionary leader who, over 30 years, has fostered a unique system of empowerment for the local women to come together. He encouraged them to not only challenge the male taboos and rules around going and fetching water but also to question the meaningless male "authority" over creating water sources in their villages for their families, and the effective management of these water sources. Sanjay Singh wanted women to be at the forefront of the work creating water sources in their parched and dry villages as it was they who bore the brunt of water's lack. And the women in the villages took up his vision! His organization formed teams that began training the women in the mechanics of creating a local water infrastructure.

BREAKING THE TABOOS

From a water-scarce area like Bundelkhand, Sri Kunwar, Phoolvati, and Neelam's stories are also the stories of hundreds of thousands of rural women across India and perhaps across the planet. Sri Kunwar, Phoolvati, and Neelam used to see the rainwater flow away. Their simple and clear thought process was that if they stopped the monsoon water then it would flow into the wells and could be used in the fields. When they were told that they should build a check dam that would hold the water for them, at first Sri Kunwar, Phoolvati, and Neelam thought, what's the point of the check dam? It only collects dirty water, not clean or pure enough to drink. But then they learned that the check dam would hold the water so that it could percolate into the ground and come up in the well as sweet, clean water.

To hold the water, they needed to build the check dam on a gentle slope so the water could run into it. They asked the men to tell them how to build all this. That knowledge wasn't the problem. They had also learned some of it watching the behavior of water and discussing it with other women when they went to the well every day. They were used to dealing with water, so it didn't take much to understand water's behavior in the landscape. They also realized that if they had the determination to hold the water, then they should be prepared to do any work related to it, so they also became the laborers doing the groundwork needed to build the check dam. The problem they faced wasn't a lack of water knowledge, but the risk of violating social norms to solve their water problems.

Through the PSSS meetings that Sri Kunwar, Phoolvati, and Neelam attended, they became aware that placing a water source nearer to their homes would mean the end of their drudgery. They continued attending meetings to receive knowledge about water-holding and what structures they needed to build to hold the water. However, because these meetings were often held in local, nearby hotels, the neighbors began to gossip, accusing them of going to hotels to meet men. Phoolvati's husband would beat her and the whole village would turn against her for being a "loose" woman. Their in-laws would get angry and shout at them for bringing shame on them.

Phoolvati told us you cannot understand how to get up until you fall down. So Neelam and the committee women, instead of moaning about their lot any more, collected money and got themselves a water distribution network of pipes in the village, even with opposition from the village men. They collected Rs.500 (60 cents) from each woman and got a pipeline laid to each of their homes with a machine for drawing water from the well to feed the pipeline. The next step was to dig the ponds for collecting rainwater and recharging the aquifers, and wells to feed the pipelines. They were determined that the water would not leave their village anymore. They would even catch the graywater coming from their homes by digging trenches around each person's house. That water would percolate into the ground and no drop would be wasted. The women had gotten into gear!

FRIENDS OF WATER TOGETHER

Phoolvati observed:

Men don't know how to cook. They don't pay any attention to what goes into it and that water has to be fetched from somewhere. No washing of clothes or cleaning up the house for the men. All they want to do is sit around in clean clothes, in clean rooms, and drink alcohol. So the women have to pay more attention to water. They say women are more bothered about and need more water but of course the truth is that men need water as much as women do.

Wherever it is dry, these solutions such as a check dam are easily available. People and farmers should not migrate from their villages, they should save the water and work on the fields in their own villages. The check dam should be high and deep enough so that one can store the maximum water. With great difficulty I also got the pipes laid for water to reach the local school as well as my home. People were suspicious why I was doing this, but I was determined that my whole village should be water-rich. There were times when they were shouting at me and threatening me, so I said to them, "Why don't you bury me with this pipe that you are not laying? I will not give up. The water belongs to all. Doesn't matter whose land it is on."

Phoolvati gets her inspiration and strength from the water and the earth. She says she knows she is not doing anything wrong. Let her village and the world say what they say. If she has a clean mind, then the world is a clean place for her. By helping construct the check dam that holds the rainwater, the level of the water in the wells near her fields has risen. They now have a hand pump in their house so she no longer has to run to the well many times a day. In Phoolvati's village, there are 11 handpumps slaking the thirst of farm animals and irrigating 400 acres.

Sri Kunwar observed:

Once the water came, I decided I wanted a kitchen garden, as too much time and money goes into shopping for vegetables which are grown with chemicals. My vegetables are fresh and good and tasty and without bad chemicals to make them grow big. I grow mangoes, tomatoes, eggplants, papayas, carrots. Then I bought a few cows. I now get milk from them and they also get lots to eat from my garden. With all this, my children can be educated. And the food they eat is fresh, tasty, and healthy.

Neelam observed:

Each village should have a river, well, pond, canal. Our ancestral water bodies should be revived and modern systems put in place alongside them. If there is water, there is a tomorrow for us and our children.

THE JAL SAHELIS

Recognizing that women are kept out of making decisions about water sources in the landscape and efforts to develop natural sources of holding water, Sanjay Singh encouraged women to form a group as friends of water, now called the Jal Sahelis.

A good start, but the first steps were not so smooth! When these women first started working to lay the water infrastructure, the men were mean and obstructive. They would give the heaviest stones to the women and watch them struggle (according to them, it was to teach them a lesson for trying to be like the men). So the women got together, shared the workload, and built the check dam themselves without the men. Of course the women get angry at the various injustices meted out to them simply for wanting to bring everyone water, but they had to live with their families and their social system too. They say they can slowly change things with the impact of their actions. There will come a time when there will be equality between men and women. In truth, it has already changed.

Women have been constrained within the male-dominated social order, and water had been controlled within the masculine systems and traditions. When these became free, everyone and everything benefited.

The men have also seen a new sense of freedom in themselves with these shifting mentalities. Now that the water has come into the village because of the efforts of the women, of course the in-laws and the husbands don't scold them anymore. In fact, all three women are called leaders in their villages. People respect what they say. This is not a superficial change of etiquette but an opening up to a transformation that has taken place in the society as a whole.

The blue saris of the Jal Sahelis show the women as flowing like the very water they hold through their work. If you gaze long enough, you might find it hard to tell the blue of the saris apart from the blue of the water. What has been held? Is it the energy of the women, or the monsoon rains that fell? What has been made apparent to the households? Is it the wisdom of the women or the ability to draw water from a hand pump within the house? What has grown as a result of this work: the confidence of the women in having their voices heard or the seeds in household gardens, now growing into food? What has been learned from this: respect and value for the women, or water being a shared responsibility of men and women?

WHAT THE WATER CHANGED

After seeing how knowledgeable and open the men at the PSSS were toward the women, the village men have shifted their attitudes toward the women and have begun to help them. Before, they would stop the women from doing water work and now they support them in it. They also join the committee meetings to plan the water collection and distribution. Now the digging is done by both men and women. Planting is done by both men and women. In fact, sometimes the men do the digging and the women do the planting, sharing the load equally.

Coming together as a band of women looking after the water has changed the landscape dramatically. When the women moved ahead, the men moved too. The women have made their voice and the voice of water heard in the world. It was only when they found their voices that hand pumps appeared. The reality they communicate is that the water was never dependent on the pipes and taps in the first place. When men and women move together like water, many things change.

There is a togetherness that wasn't there before. Small nurseries have opened for the children. Before, their families never ate together—the men would eat first and then the women would eat whatever was left. But now, each family will often sit for meals together. As respect for women has increased in the society, so has the men's own self-respect, upending centuries of mindless male authoritarianism. Many politicians come and visit to applaud the work in regions that previously most people hadn't heard about, never mind being beneficiaries of government attention.

So the change that speaking the language of water brought about was in easing the centuries of societal separation between the women and the men!

Neelam noticed:

> It was very hard to change the men and society but they have now understood. We managed to get it out of their heads and out of society's heads that water is only women's work. The men also learned and took the brave step to go beyond the social norms. The women kept asking and, at some point, the men relented and now we reap the benefits together.

WHAT THE WATER TEACHES

The water gave strength to both the men and the women to do the right thing and change. The softness of water permeated the men until they melted and gave the women the freedom to do what needed to be done. And, like water, the women got the strength to go ahead and do what was needed in spite of all the odds they faced. Women enabled the men to become strong inside and move toward empathy and equality and the men enabled the women to be strong outside and bring back the water.

"No matter what troubles you face, make your villages water rich. You will move ahead and when women move ahead, then the whole of society moves ahead." –Sanjay Singh.

The Jal Sahelis now number 1,500 and growing, a strong group of beautiful, courageous, powerful rural women in the region of

Bundelkhand in central India—one of the driest parts of the country. Wearing their beautiful blue saris and bright smiles, they say they represent that life-giving drop of water that has kept their families and villages alive.

This growing group of women is responsible for carrying forward the water security agenda of their regions and providing leadership toward a collective assertion of rights. They are responsible for creating awareness among their communities about water rights and accessing various government programs to get flowing water. PSSS and the Jal Sahelis started village level, community-based water committees working for protection, conservation, and management of water resources. They are also providing leadership for claiming entitlements and rights for water that are due them, as well as strengthening local self-governance under the strong leadership of women and marginalized groups. "Pani panchayats are necessary to fight the battle for water," says Sanjay Singh.

Each pani panchayat has 20 to 30 members, and now there are both men and women involved. These members nominate four or five women as Jal Sahelis, who drive the implementation of water-related works in their village. Pani panchayats draw up the water security plan for the village including the total availability of water and its requirements.

The Jal Sahelis are responsible for raising water issues at the village level and mobilize their community to participate in village water development as well as in the preparation of a water user master plan for their village. Jal Sahelis provide solutions to all kinds of problems in their villages, including water harvesting, water conservation, deepening wells, rehabilitating water structures, building small dams, improving hand pumps, getting community participation with the government, meeting administrative officials, and submitting memorandums.

The local Jal Saheli groups now have rich experience in improved and sustainable agricultural practices, promoting efficient water use, water and soil management, establishing water–livelihood linkages, watershed management, policy, advocacy, networking, and alliance building, which they share widely at their village level.

Through the pani panchayat, these Jal Sahelis have provided a total of 10 million cubic meters of water to 2,256 hectares of agricultural land. A total of 3.288 million cubic meters of water was saved through smart changes in cropping patterns and agriculture. Due to these efforts, their regions have seen an additional agricultural production of 1194.60 tons and employment of 8,867 laborers.

"The first priority is drinking water, followed by sanitation, gray water management, and irrigation," Neelam says. After this, Jal Sahelis ensure the implementation of the pani panchayat's plan.

Today, Sri Kunwar, Neelam, and Phoolvati are in their 30s. They receive great respect from their community and are treated like elders. Older women in the villages take inspiration and join the Jal Saheli program, saying there is no age restriction for receiving an education.

Some of the women who weren't educated when they were young are now working with water and bringing water into their villages. They have even learned to sign documents with a pen as opposed to just a thumbprint. Such is the change that water has brought!

The Jal Saheli and pani panchayat programs of PSSS are the most successful community models of water revival in the country and even received an award from the president of India in 2023. The strength of the Jal Saheli movement is in the realization that the social restrictions on women's place within the family went hand in hand with communication around the management of water, which was also prejudiced in favor of large engineering solutions rather than natural, community-led work.

This brings us to the inner heart of it all. Is the language of water feminine or masculine? Usually water is considered feminine. So is the earth. In so many places, it's the women who provide the food and water for their families. In rural locations, they are the ones who walk miles to get the water, often carrying this heavy load on their heads or on those of their daughters. When you go on an international trip to an exotic destination, you are often allowed to check in a 23-kilogram bag. Imagine carrying that weight on your head instead! That's what women in rural locations have to carry to fetch water for their families, many times a day.

Yet for all that, they are the ones who are kept out of the decisions that are made around the sources of water. Common sense says that those who suffer make the best decisions to help end their suffering. Not so with water. In many patriarchal cultures, it is the men who decide what to do about the water, or lack of it, in their communities. In patriarchal and conservative traditional cultures in certain parts of India, women aren't allowed out of their homes unless accompanied by a male guardian, yet ironically, when it comes to the hard graft of having to go out and fetch water from a dirty well or pond miles away, miraculously that restriction is waived and off they are sent to bring back that water on their heads for the family to use. The lesson our three Jal Sahelis give us is that the rigidity that had developed around the use of water was the same as prejudice around the place of women in society. Women need to take that first step to change it.

Phoolvati says:

> If we as women are not aware and awake, no one will listen to us. We have to become aware and awake and then the whole of society will come along with us. Nobody else will do it for us. We have to do it for ourselves. The way I am making my village water rich and water secure is the same way the country, the world, and the whole planet can revive. If men and women unite and work, educate both boys and girls, treat them equally, we can make the barren land lush. Why leave it barren?
>
> It will help the planet to revive it. I really wanted to study, but poverty stopped my school education. Now I am educating the whole village. There is education in the books and education of the heart and the spirit. We need both, but if you can't have both then the latter is what you must have. What we have taught the world is far more than many educated people.

- Where there is barren land—hold the water.
- Look where the water comes to a halt and hold it there.
- Wake the women.
- Wake society for men and women to work together.
- The qualities that women have are also those of water—increase those qualities.
- The next generation should not suffer what we have suffered.

- Educate the children. Education opens the mind and sorts the problems in the world.

Sri Kunwar says:

> Water is definitely feminine. If there is no water, then there is nothing. If there are no women then there is no life. Water is used in everything. Just as we say Mother Earth, what will we say for water? Mother Water or Father Water? For water, we can only say mother. Because only the kind of love and looking after that a mother can give to a child is what water gives to us. Women use water and know how to use it wisely. When there is less of it, they know how to stretch it. Like a mother gives birth to life, so does water give us life.

Neelam says:

> I have three children: two daughters and one son. All of them are studying and my eldest daughter is now enrolled in a graduate degree course and dreams of joining the police force! No child marriage for her and she is free to study and follow all her dreams! My mother-in-law says that the children must complete their education. She feels so glad that this water work has been done by me, her daughter-in-law, and above all she really appreciates the togetherness between the men and the women that the solution of the water has brought to our village.

The women could not have done the work on determination alone; Sanjay could not have had a vision without the determination of these women to change their lot. Thousands of years of male dominance is transforming into equality for rural women in India. A rigid system with strict patriarchal rules slowly opening up and beginning to flow like water. The language of water is like the yin and the yang. Only when both the men and women speak it together does the water cycle return to the region.

A poem written by Neelam Jha, *Jal Saheli*:

मैं जल सहेली नहीं
मैं जल सहेली नहीं
मैं एक पानी की बूंद हूँ
इस बूंद से बहुत बड़ा समुद्र बनाऊंगी
अपने बुंदेलखंड को ही नहीं
पूरे भारत और पृथ्वी को जल संकट मुक्त करआऊंगी
मैं जल सहेली नहीं एक पानी की बूंद हूँ।

(translation)

I am not just a friend of water
I am like a drop of water
Like a drop of water I will fill a vast ocean
An ocean that will benefit not just Bundelkhand
I will fill an ocean that will make the whole of India water rich
I am not just a friend of water,
I am the drop of water that fills the ocean to make the world water rich.

14

United States/United Kingdom/Europe: Beavers, Nature's Water Engineers

The beavers, should they be given rights, would own part of the copyright of this book on the language of water. (After all, we are using all the royalties from this book to build beaver-like water-retention projects!) Beavers have proved, by their absence and reintroduction in the wild, to be an exceptional asset in making a balanced water landscape. Even so, in the past they have been seen as mere pests that ruin good agricultural land. Hunted in the United Kingdom throughout the 1500s for their fur and oil, they were driven to extinction by the beginning of the 16th century. When wild beavers were spotted in this century in Devon, United Kingdom, it led to the establishment of the River Otter Beaver Trial and other research projects to examine whether they were pests or not. Visiting the River Otter Beaver Trial, we immediately fell in love with the animals that so comprehensively transform a monotonous, grassy field landscape into an aquatic home, embodying all the principles of holding water that communities emulate to become water-rich. We were eager to research the history of what they did, where they came from, and what happened to nature's water engineering community.

WHY A BEAVER?

Beaver landscape; Yettington dam from air. Courtesy of Devon Wildlife Trust.

A beaver. Courtesy of Michael Symes Devon Wildlife Trust.

A beaver is a completely different type of rodent from the mouse that scuttles shyly through the undergrowth in search of a good hiding place. For a start, a beaver is much bigger, weighing 40 to 60 pounds as an adult. Second, it hides from predators by transforming its landscape to a semiaquatic habitat in which it can naturally outswim its foes. Boldly the beaver goes about creating this diverse and varied habitat. Instead of building a nest in a tree or digging a hole in the ground for rearing their offspring, the natural expression of beavers is to establish significant wetland areas, using a mixture of tree felling (something like leaky log dams?), hydrological engineering, canal digging, and dam building (something like *johads*?), that afford it vast and ample space to live a healthy, abundant, and luxurious life.

Beavers are nocturnal, working at night going about their business of systematically holding water and transforming their living spaces. They are similar to humans in that they seek to collaborate with a living region that supports a healthy water cycle. As Ben Goldfarb says in his bestselling book *Eager*, "Beavers have permanently shaped the course of biology and geology. They have re-formed rivers, kneaded meadows, filled valleys. We built our civilization atop the sediment they left behind."[1]

To do this, the beaver is equipped with:

- reinforced teeth strong enough to gnaw through trees
- a set of transparent eyelids as a natural pair of goggles to see underwater
- a second set of lips behind their front teeth enabling them to chew and carry branches underwater without drowning
- paws that can dig out a channel network of irrigation ditches, off from the main dam, enhancing the wetlands
- a tail that can act as a rudder
- webbed, ducklike hind feet for swimming
- a dense fur wetsuit
- a social, peaceful, and of course eager and hardworking nature

Instead of scurrying about as a pigeon does, clumsily holding an odd, dry twig in its beak, beavers work ceaselessly to transform a whole field-based local environment into a lush wetland. The beaver creates a whole different ecology, not to protect itself in a den into which no other animal can enter, but to share a habitat in which it is especially suited to thrive and that allows other life to flourish as well.

HYDROLOGISTS OR PESTS?

Beaver handiwork. Photo courtesy of The Flow Partnership.

Visiting friends in Scotland in 2014, we went for a walk near Loch Coil-le-Bharr, where the Knapdale Beaver Project[2] had brought beavers in from Norway in 2009 as part of the Scottish Beaver Trial.[3] All the negative publicity about pests and rodents and the controversy of the trial was on our minds, so we were expecting to see some clumsy intruder making some token impression upon the landscape, destructive of agricultural aims. We did not pay much attention to the site or even understand the vast hydrological engineering that was on display. It all just looked like one messy eyesore to us then.

It was only when we began researching the modus operandi of beavers and the impact of their actions on the landscape that it began to dawn on us that they were truly the original model of nature's engineers and not the pests they were purported to be. They are more like ancient guardians and speakers of the language of water. They shape the most exquisite environments by creating new expressions of water, land, and vegetation and teach us humans that language in the process too.

REGISTERING IN A BEAVER CLASS

Hundreds of years ago, in the 16th through the 18th centuries, beavers became extinct in the United Kingdom and in almost all of Europe. They had been overhunted for their fur, meat, and castoreum (a secretion from their castor sacs, used in perfumes, food, and medicine). No one at that time realized beavers' vital role in the ecosystem and maintaining the landscape's water balance. It is symbolic of the disregard for the language of water that its natural speakers were hunted to extinction.

In 2008, something miraculous happened in the United Kingdom. Beavers of unknown origin were spotted in the wild on the River Otter in Devon. However, no one was sure it was really beavers they had spotted. In 2014, when videos emerged of them successfully giving birth to young, the beavers entered the public spotlight. Much to the consternation of the UK government and the delight of naturalists, headlines about the beavers began to appear in various news media.

The Guardian reported: "Wild beavers seen in England for first time in centuries: A family of wild beavers has been seen in the English countryside in what is believed to be the first sighting of its kind in up to 500 years."[4]

The BBC reported: "A spokesman for the Department for Food, Environment and Rural Affairs said it was unlawful to release beavers in England and they were looking into what action to take."[5]

In response to the sighting, DEFRA (the UK Department for Environment Food and Rural Affairs) voiced their concerns and advised that the beavers be removed: "Since the presence of the pair of beavers on the River Otter was made known to us earlier this year, we have been concerned for two reasons [their unlicensed state and their potential for hosting disease]."[6]

However, the government was persuaded to send the beaver pair that was spotted in the wild to be housed for a trial at the River Otter Beaver Trial [Devon Wildlife Trust] site near Exeter. Started in 2015, the site was supported by Natural England (a UK government advisory body for the natural environment) over a period of five years to monitor the effects that beavers have on the environment.[7]

In this River Otter Beaver Trial, a few fields of grass—an ecologically degraded and bare area—were fenced off and the pair of beavers were let loose on it. Fast forward a few months and lo and behold! Those same bare, inert fields had been transformed into healthy, lush wetland, supportive of diverse life. Researchers and scientists from the Devon Wildlife Trust and Exeter University monitored the volume and quality of water entering and leaving the enclosed trial site in order to understand the impact that beaver dams could have on flood risk downstream and diffuse pollutants in the watercourse. The results have been dramatically positive to say the least, and hugely beneficial for a wide range of wetland plants and animals.[8]

CAN ANYONE SPOT THEM?

In 2018, we took a group on a Water Summer School visit to the River Otter Beaver Trial site. This time we paid attention to the wonder of their work. We spent hours walking across the site but didn't see a single beaver, of course, as they are nocturnal creatures and sleep during the day. Since they mostly work at night, when you visit a beaver site, you wonder what sort of community has been responsible for such a transformation of the landscape. The dull fields you trudge through to enter the site suddenly transform into an intricate aquatic environment, superbly designed and brought to fruition through a complex series of structures, dams, and interventions. Surely some master engineers have been hard at work in here!

We could see ample evidence of the beavers' work through the fallen trees and odd bits of shrubbery among land, wetland, and green-ery—lots of water being held in different ways to create the environ-ment the beavers needed to flourish. Beavers are supremely attuned to their environment and work via deliberate design and strategic collec-tive work. When the rains come, the standard of the beavers' work is advanced enough to slow the flow and reduce the chance of flooding downstream. If the conditions dry out, then the beavers have succeeded in storing tens of thousands of cubic meters of rainfall into the ponds, structures, and underground aquifers. There is nothing makeshift about what the beavers do.

WHO IS FLUENT IN THE LANGUAGE OF WATER?

The language of water can be understood as a useful reinterpretation of the laws of flow as a potentially groundbreaking new solution to the cri-ses of floods and droughts. In fact, the language of water is not a human language at all. It is spoken by nature itself and, in the work of the bea-vers, we see its expression in the harmony among the elements: earth, air, water, sky, fire, and vegetation. The scientists guiding us through the area marvel at the beavers' ability to invent without the least need for intellectual training.

What we mean by a language of water is not a human construct. The beavers speak that language by attending to every detail of the inter-ventions needed to sustain the aquatic system in collaboration with the elements, illuminating nature as a series of complementary languages articulating the potential in a world we inhabit together—a world whose tendency is toward transformation, as in the water cycle.

The human language of theoretical understanding, which imagines it is the only system for lending meaning to the world, has shown itself, in the practical realm of tending to the environment, as less able than the beavers. The beavers seem to understand the transformation of the land all at once and attend to the details accordingly, whereas humans tend to identify the details and then try to fit them into a whole design. The beaver articulates the potential to relate to the world with a spirit of transformation—a language innately available to us humans too. We have only lost our familiarity because we are no longer in the practice of it. It is as simple as coming back to that practice again.

TURNING THE KEY TO TRANSFORMATION

Faced with the climate challenges of extreme weather events, droughts and floods becoming the norm, and water running off instead of being held in the land, the human intellect is distracted by measurements and huge visions of engineering construction, addressing these as individual challenges and isolated projects. The beaver, on the other hand, faces ecological challenges with the potential of its craft, to shape a whole solution entirely in keeping with the capacity of the environment to support life.

Witnessing the beaver's inventiveness in creating a suitable habitat for itself, the only answer seems to be that beaver and wetland landscape arise together. The language of water emerges to express imaginative solutions within the landscape before the intellect comes along to measure and analyze what is going on. This might seem like fanciful conjecture, but the beaver has not been to engineering school and yet it solves the problem of flooding downstream by slowing the water on the land it occupies.

Successfully creating a wetland, as the beaver does, involves synchronizing multiple tasks such as felling trees and building dams to strategically intercept watercourses. Human intelligence (for example, debating whether to allow the beavers to return to the River Otter) can seem cumbersome when placed alongside the communal work of the beavers. It is within the human imagination to access the innate skills of the beaver. Where a human-engineered wetland might boast a few willows and the odd pond (and volumes of unrelated data, policy, and reports), the beaver's work results in a completely altered environment, as if one has entered a quite different world. When humans trust an innate sense of nature's regenerative order, then we too can be a part of the unity of water, land, and forest. To know the beaver is to understand some part of ourselves in the ability to bring about creative change at a whole scale.

IN FOOLISH PURSUIT OF A HAT

Unfortunately, it is not just the United Kingdom and Europe that witnessed a steep decline in beaver populations from overtrapping. North America witnessed the same trend during the same era. According to Butler and Malanson's research in a 2005 issue of *Geomorphology*:

> Removal of beavers by overtrapping in the 16th–19th centuries severely reduced their number and the number of ponds and dams. Dam removal altered the fluvial landscape of North America, inducing sediment evacuation and entrenchment in concert with widespread reduction in the wetlands environments.[9]

The extinction was driven by the trade in hats made from beaver fur between the 17th and the 19th centuries. How short-sighted is it that to protect the brain—seemingly the source of human cleverness—the natural speakers of the language of water were pushed to near extinction. Butler and Malanson also tell us that:

> Uncounted millions of beaver ponds and dams existed in North America prior to European contact and colonization. These ponds acted as sediment traps that contained tens to hundreds of billions of cubic meters of sediment that would otherwise have passed through the fluvial system.

The beaver wetland environments were a formative part of the landscape, embedding water and its dissolved sediment in the land in parallel with the water cycle of the trees evaporating the water into the atmosphere. How many communities would it take to create the millions of ponds that the beavers can make? The life and sediment that existed in those ponds must have translated to billions of tons of carbon sequestered per year, which are now being released into the atmosphere as waste CO^2, adding to human-induced climate change.

Beaver dams are not perfect; research demonstrates fatalities from outburst floods when their dams collapse. But human dams also fail or collapse, one recent example being the collapse of the Derna dams in Libya in September of 2023, which killed tens of thousands. Beavers cannot be judged for the imperfections of their art when scientific communities have created such instability in the climate and landscapes globally by following their own laws.

According to Butler and Malanson, in the 19th century, flush with the Darwinian conception of nature as a brutal battleground of survival, which only human calculation could oversee, the population of beavers dropped to 100,000 in North America and 1,200 in Europe.

> Partial recovery of beaver populations in the 20th century has allowed reoccupation of the entirety of the pre-contact range, but at densities of only one-tenth the numbers. Nevertheless, modern beaver ponds also trap large volumes of sediment in the high hundred millions to low billions of cubic meters range.[10]

Beaver population was estimated in 2021 at 6–12 million beavers in the United States, compared with 60–400 million in their heyday.[11] Of course, modern engineering methods can be part of holding water in the landscape, helping to deal with extreme climate events. But modern engineering would also do well to learn to mimic the intricate and holistic stylings of the beaver, speaking the language of water in concert with the land.

TRIBUTE TO THE BEAVERS

Beavers are a vital link in the environmental landscape. They show us that integrating water into the land is not a one-off wonder that can be accomplished overnight—transforming the landscape to hold water is hard and necessary work done patiently and holistically.

The Wild Trout Trust supports beavers for the beneficial impact their dams have on wild trout and other fish:

Where beavers build dams, water will be stored in the wetlands and ponds behind the dams. The ponds fill to the depth of water dictated by the height of the beaver dam. Beavers are rather efficient at plugging leaks in dams to maintain water levels (to the 0.6m depth they prefer), so when the ponds and wetlands reach their storage capacity (dictated by the beaver dam height), that's it—any excess water can percolate through a leaky dam or overtop it and carry on down the stream.[12]

Beavers teach us to establish significant wetland areas using a mixture of tree felling, hydrological engineering, canal digging, and dam building—holding water where it falls in the landscape.

The beaver:

- strategically gnaws through and fells trees (so we, too, can create leaky log dams along flow pathways to slow the flow of water rushing downstream);
- creates a dam and a series of 15–20 sub-dams to hold the water (so we, too, can create multiple water-holding structures to infiltrate and recharge underground aquifers);
- uses innate hydrological acumen to place the dam structures effectively in flow pathways (so we, too, can create structures along flow pathways where the maximum amount of water accumulates, creating the potential for maximum water holding);
- digs out a supporting canal system to distribute the water widely over the surrounding areas, stopping droughts and floods (so we, too, can contend with droughts and floods by distributing water evenly over a large area).

These principles will stand up to the test of nature wherever they are applied on the planet to reduce floods and droughts and lessen the impacts of climate change, human induced or otherwise.

LANDMARK DECISION GIVES WILD BEAVERS PERMANENT RIGHT TO REMAIN IN ENGLAND

DEFRA, after considering the report of the River Otter Beaver Trial, gave the go-ahead for the reintroduction of the beavers into the wild:

> Today, a landmark decision from Defra announced the decision for these enigmatic mammals to be given a permanent right to remain in their East Devon river home, securing their future in England.
>
> The Government made its decision for the mammals to permanently remain on the River Otter thanks to evidence in a report published earlier this year from the River Otter Beaver Trial (ROBT).
>
> The ROBT—run by Devon Wildlife Trust and partners—concluded that the animals' presence was overwhelmingly beneficial to the people and wildlife living along East Devon's River Otter.[13]

Now there are plenty of beavers all over the United Kingdom with the blessings of the government; some are even found in the wild, not just on trial sites, with a crucial difference—this time around it is not for their fur or oil that they have created such a huge stir and excitement. It is for their understanding of the language of water, spoken so effectively in the landscape. This time, we're listening.

Colombia, South America: Steps to Community Action

People do want something to be done about climate change. They see how dirty their local river or landscape has become and wrinkle their noses in disgust, saying, "Someone should do something about this!" Often, they rely on their governments to do that "something" and sit back from their own responsibilities and, in that, do nothing much at all.

Who will become the someone who finally does something about the degrading state of a local water system? What is that dirty creek, the only source of fresh water flowing through the town, telling them? Can people heed its message beyond the disgust and come together as a community, taking action not only to clean it but to benefit from it as a source of joy, pleasure, and health, for now and for future generations? Perhaps the hardest part of any project is taking those first steps together as a community to make that change. Where do you even begin?

No matter how aligned the community is in their vision, purpose, aims, or disgust at the state of affairs, without action, nothing will change. What is that first step one can take? What informs that impulse to action? What speaks to the community to find it in themselves to set aside the vested interests, the hopelessness, the frustrations, and take action together?

As an example of what can happen when these first, brave steps are taken, we have been following a creek restoration project that is currently unfolding in Barichara, Colombia, a small town of 7,000 inhabitants, located in the northeast of the country in a tropical dry forest ecosystem.

Extending back to the 10th century, the original Indigenous inhabitants of Barichara were the Guane people. As with all Indigenous people, they understood that their survival was directly mediated by their relationship with nature.

In the epochs of civilization that replaced the Indigenous people, there followed a progressively more outrageous bet on whether humanity needed nature at all. In each epoch, when nature demonstrated the consequences of neglect, a harsher regime of measures was brought in to further control it.

During the Green Revolution of the 1960s, tobacco monocrops in Barichara altered the fragile climatic equilibrium of the waters. Over that period, more than 80 percent of the tributaries that nourished Barichara's stream ran dry or became seasonal creeks, creating a water crisis that exists to this day. Today, during the dry season, there are constant water shortages, often every week.

When we feel concern for the effects of climate change happening right at our doorstep, it is helpful to see how we arrived at this situation as the result of incremental steps pushing nature further and further away. When we understand that we are all implicated in this distancing from nature, an opportunity opens up for us as a community to change our local environment.

In May 2020, during the first phase of the COVID crisis, road blockages and lockdown restrictions affected the town of Barichara severely, isolated as it is by its geographical remoteness. An exponential rise in the cost of groceries coincided with restrictions that were imposed on the export of local produce. The scene was set for a long-sought-after dream of creating a local, shared healthy food distribution system.

Initially 12 families got together—six buyer families and six farmer-producers—and bypassed all legalities to form a small food co-op, claiming a higher order of responsibility to care for their lives. The co-op grew at a marvelous speed and by week eight they had approximately 14 producer families and 20–30 buyer families in the co-op.

Suddenly, in mid-April, one of the farmers called the co-op members before harvest day. "Don Carlos," she said to the partner, "I'm sorry but tomorrow I cannot send you any food. The last drop of water on my *jaguey* (body of water) has dried up."

Felipe Medina, a resident and father of four children in Baricahara, recounts that this moment planted an urgent intention in him to find a creative and sustainable way forward through their water crisis.

STIMULATING ACTION

Having heard of our work with communities in Rajasthan, India, Felipe Medina contacted The Flow Partnership with tales of these dry tributaries in Barichara, wondering what they could do and how they might involve their community. We discussed how they might create the motivation that could bring about transformative change.

The story that Felipe told of Barichara is not exceptional. It traces how a once-thriving ecosystem, through neglect and miscalculation, has been turned into a water-deprived area without a working waste disposal treatment plant, reliant on a faraway reservoir to pump water in from outside the region. The traditional methods that used to hold the water have been disregarded, the creeks and springs are dry, and water shortages are now common. The forest has been cut, the land has dried out, and the rainfall has diminished without the trees acting to circulate the moisture in the air. Abundant water on our planet is not a guarantee. There is nothing to stop us from breaking the water cycle anywhere and everywhere if we choose to do so. Through our actions, we can easily, over time, become a dry and dusty planet like Mars, incapable of supporting life. Water presents the same challenge for Felipe Medina[1] in Colombia as it does for Rajendra Singh in India[2] or Zephaniah Phiri in Africa.[3]

Felipe was known as a proponent of nonviolent communication. He met Claudia Steiner,[4] an ex-teacher, at a meeting in Barichara where he was mediating a difficult discussion between a builder and some environmentalists. Impressed with his calm manner as well as his vision, Claudia asked Felipe to propose a project that would help address the degrading ecosystem in Barichara, offering to fund whatever plan he devised. This led to the Senderos del Agua/Water Trails project, a children's pilgrimage and research initiative designed to stoke the memory and revive the imagination of water.[5]

This project was further supported by three elders in the community: Camila Encinales, María Victoria Camacho, and Alberto Botero, who in 2011 had begun working toward a Green Belt in Barichara, an ecological corridor that aimed to surround the whole town. These elders welcomed the Senderos del Agua/Water Trails project and the initiative to engage the community by walking along the creeks in the Green Belt.

THE APPROACH

The water crisis in the co-op showed that the area could not be self-sufficient in food unless they tackled the problem of water first. This prompted the Senderos del Agua/Water Trails project to seek the origins of the crisis within the memory of the people of Barichara. The project began by interviewing many community members and reconstructing the history of how the present water crisis had begun. It also asked the question, using a community empowerment approach, of how the people would like to be transformed along with their water in the present day. First, they outlined and agreed to the purpose and principles that would guide their work:

Purpose

1. To connect memory and imagination around the local water.
2. To generate a common understanding of the historical events that had led to the community losing the ample water that used to run through the region.
3. To promote a connection between the different stakeholders from a position of common responsibility.
4. To strengthen community empowerment based on these connections and understanding in order to constructively influence water regeneration.

Principles of Action

1. Put water at the center: it is the source of all life.
2. Be guided by the children: the living possibility of the next generations after us.

3. Walk without creating enemies—we are all responsible for water, its current state and its regeneration.

4. Listen—to ourselves, to others, to the territory—and trust that the answers will emerge.

5. The starting point of water regeneration begins with ourselves.

Actions

1. To research the historical reasons for the malaise of the creek and the constant water shortages.

2. To walk along the creek with the community to regain ownership of the stream that had been made part of a much larger water provision system.

3. To renew the natural flow of the water.

Senderos del Agua/Water Trails organized three pilgrimages on three- to five-kilometer stretches of land along the creek and its tributaries in the Green Belt. Before and during each walk, interviews were carried out with everyone who had lived on that stretch, who owned land, or had a connection to that place. A cross-section of the community—from the priest to the fireman to the water department to the car washers—was interviewed, all of whom offered hospitality and support. When asking permission for the children to walk their stretch, the owners of the land were most welcoming, cleaning the route and treating them as honored guests.

The Senderos del Agua/Water Trails project was first and foremost a vision that belonged to the children. It arose from their will to revitalize the Barichara stream. It harbored their dream that its water would run so clean that it could be drunk. The project offers a solution to the region's long-standing water crisis by restoring the surviving stream that runs right through its center. It was an invitation to ask, what would happen if—when imagining future possibilities—we really listened to the children?

The project was also novel in that it gave space to speak to the community without imposing a pre-assumed narrative or suggesting a solution. As they walked with their children, they asked about the past, present, and future of water, exploring how their relationship with water had been transforming, holding space for visions of a new future. The project sought the reasons in the past for the degradation of the creeks and found the answer in the children's relationship to their own future.

The children in a project exercise. Photo courtesy of Felipe Medina.

During the walk, the research group shared stories they had collected from the elders of when the creeks were still healthy to compare with the foul-smelling water they encountered. Childhood stories, speculations, gossip, popular knowledge, semi-mythological accounts, and detailed narratives with chronology were woven together to compose a kaleidoscope of perspectives in the form of a living history—a kind of polyphonic song of human experience of the territory. On this basis, to construct a narrative, they identified the common patterns and the most significant subjective features of the different accounts. These insights sought to locate not only what had happened but also the deep ecological and cultural forces that had shaped the past and could further shape the transformation of the territory.

A TENTATIVE STEP

Children taking part in the Senderos del Agua/Water Trails Project. Photo courtesy of Felipe Medina.

In December 2019 the project did the first pilgrimage, walking the territory with 14 children around Barichara's Green Belt. The place where the old spring was (on the now mainly dry Quebrada la Vieja tributary to the Barichara Creek) still holds moisture right in the middle of summer. As the group arrived there, the children started saying, "I hear water, can you hear it?" But when they reached it, they found the majestic Barichara Creek turned into a horrible, filthy sewer. Rancid and foul smells filled the air. The group sat on the grass and drew up possibilities about what should be done. And then, slowly, the stench of the current sad state of the creek was overwritten with the imagination of the children.

"As I toss these rocks into that dry grass, it sounds like puddles, can you hear it?" "If I put my ear near this tree, I hear a dripping sound, can you hear it?" Finally, as they went to place a small offering between the first dam and the little spring, the youngest girl whispered to the group, "The Guane [the Indigenous people who used to live here] just told me that we saved the snake [symbolizing fertility and water), did you hear it?"[6] The old mythology of the Guane came alive in the observation of the child.

The secretary of culture of Barichara, when shown a color-coded water map of the route made for the children, responded, "Right in this spot there was plenty of running water. I am remembering now. When I was a child there was an elder woman who would look out for me. I needed to cross this creek every day to go to her house. I would look at the fish and play in the water. I had completely forgotten about this. I am remembering now. Now I will not forget."

One of the people who came on that first walk was Joe Brewer, an atmospheric scientist, who along with his wife had moved to Barichara from the United States seeking ways to contribute to ecological restoration in the area. Writing after the walk, Joe pays tribute to the impact it had on his whole family:

> We were welcomed by a host family—Felipe and Alejandra along with their three young kids—who were about to begin a summer school for children based on the principles of deep ecology. Our daughter Elise joined the boisterous group of twelve kids ranging in age from three to seven on a walking journey around the ecological corridor surrounding this town. Our goals of learning how to live regeneratively remain largely elusive. . . . this deep [dream] of a life transformation takes time to manifest, after all. . . . But it wasn't until we walked in the forest with a group of children at play that we began to actually do something about it. And this small step may prove to be all the difference in our lives.[7]

CONVERSING WITH THE GOVERNMENT

By June 2020, the project leaders began conversations with the local government about the possibility of cleaning up Barichara Creek and specifically about fixing the town's water treatment plant, which had been broken for years and was discharging its black water directly into the creek's basin. They were met with bureaucracy standing in their path, an apparently unmovable obstacle. The government said they could not intervene in the water situation because of legal wrangling that had first to be resolved by Colombian law. No new projects could start until these issues were resolved. This could take years.

There were two disputes around allocated funds failing to deliver the necessary upgrades to the water: First, 2 billion Colombian pesos ($500,000) had been lost due to the incomplete delivery of the municipality's Waste Water Treatment Plant (WWTP). These resources were allegedly invested in the work, but due to corruption, the WWTP is still yet to become operational.[8]

Second, Jonathan Malagón, the minister of housing, referred to legal issues that prevented a new aqueduct from solving the water shortages in Barichara. A project had been approved in 2012 but was prevented by the objections of the municipality of Galán, from where the water was being taken. The project stalled and the money that had been allocated disappeared.[9]

How did this impasse come about?

THE PAST INFORMS THE PRESENT

Far from pretending to be an "official truth" or an "objective" account, this summary of the historical and cultural movements seeks rather to answer these questions through the community. What happened to the water? How do they make it come back? What are the opportunities at this moment? The foundation of this work is based on trust, which lies not in the extrinsic value of the narrative but in the value it may have in providing a coherent proposal of alternatives for sustainable water management into the future.

One of the aims of the project was to document the history of the area and how this gradual schism between the land and the community had arisen. The history that was discovered is replicated in detail here, as it echoes how colonialism around the world overwrote the original collaboration between culture and nature. A succession of five historical moments related to the transformation of the cultural and ecosystemic reality of water and territory underpins the state today.

PRECOLONIALISM

The original inhabitants of Barichara, before colonization in the 16th century, were the Guane people. As farmers of cotton and skilled artisans of cotton textiles, they traded, especially with their neighbors. Their mythology would have been closely connected with their natural territory.

COLONIALISM

In colonization, one can see the gradual desacralization of the land. Cultural historian Thomas Berry makes a clear point here: before

Catholicism, most cultures had a territorially codified relationship with what they understood as sacred.[10] That is, the territory they lived in was "the book" from which they understood and derived the meaning of life and its transcendent aspects. With the arrival of Catholicism in the occupied lands, this relationship ceased to be anchored in the specific reality of a territory and became more about referring to the contents and enunciations of the Bible.

This desacralization of the land surely had strong implications in the people's relationship with water. The use of the streets, and later the rivers, as a mechanism for the discharge of sewage started during this colonial period. Although no specific records were found, reports from local historian Juan Carlos Parda say that the Barichara sewage system was one of the "oldest and best built" in the region, with stone vaults that led the sewage from the houses directly to the stream.

From this point on, we can imagine that a decisive transformation began to take place in the people's relationship to the land, a transformation that survives to this day. The concepts of usufruct and extraction became the main cultural determinants in the territory, replacing the notions of value and sacredness that had been predominant in Indigenous times.

MODERN TIMES

In Barichara, "modernity" (referring to an era of technological, scientific, and socioeconomic progress) arrived only fairly recently. Fifty years ago, there were only two cars in Barichara, and some of the roads were still dirt tracks. The economy was based on agriculture. People were still living with a direct relationship to water in a very similar way to their ancestors. Local people drew their water from the many cisterns or springs found in the region. From those springs, of which there were as many as 83 located in the urban center, water was distributed to the entire population. Those springs in turn were supplied by underground streams, which were sourced directly from the forests and other natural water recharge systems in the region.

For millennia, water had depended on the quality of the land and the people tending it. A "good" relationship with neighbors in the treatment and care of water was a matter of survival, reflecting the popular wisdom that "When people fight over water, the cistern dries up."[11] The governing principle behind this popular belief, widely shared in the region, is that of radical interdependence between human and ecological systems. At this point, water was understood as a common good and a collective responsibility, whose ultimate care depended on human relations in the territory. This history changed with the arrival of the aqueduct.

The current malaise in the water situation in Barichara began when water was brought in through the artificial aqueduct system in 1981 through reservoirs and pipes run by the water company. The advent of a modern system displaced the trust in traditional measures of holding water, and the local springs and creeks were allowed to dry up or be filled with sewage, as they were no longer deemed to be functionally important.

With the arrival of the aqueduct, the responsibility for the supply and management of water had now shifted from the hands of the local communities to those of the state. The cisterns that formerly existed in households were replaced by the modern aqueduct channeling water from afar. The water supply no longer depended directly on the local ecosystemic composition of the territory. The responsibility had been transferred to a distant region with which the community had no immediate relationship. When the town was severed from the local water supply, Barichara surrendered its water self-sufficiency and succumbed to water shortages and crises.

The possibility of profiting at the expense of others became the prevailing logic in the territory's water management. The cultural shifts that arose with the arrival of the aqueduct coincided with two other closely related phenomena. The first was the arrival of the Green Revolution, which in some way represented a similar pattern of change in terms of water but now in relation to the land. The Green Revolution, with the logic and technologies it offered, came with the message of turning the land into a machine of sorts, capable of supplying produce at an accelerated rate, denying the ecosystemic processes of self-regulation and regeneration on which all agriculture previously depended, and offering industrial alternatives that were costly and harmful for ecological and human systems in the medium and long term.

This moment also coincided with an era of partisan violence in the country and particularly in the territory of Barichara. This violence brought with it the fracturing of relations and the second forced displacement of one community by another for economic purposes around the control of the territory. In this case, it was the Liberals who were forcibly displaced from their lands, and the Conservatives who in many cases occupied their sites. Currently, we are beginning to see litigation that seeks to vindicate the forced displacement of that time period, with families who were displaced returning to reclaim their land.

It was around this time that the 14 tributaries that had fed Barichara until approximately 50 years ago began to dry up one by one. Today, only one, Fiques Creek, remains with flowing water. If the combined flow of these 14 streams were added together, it almost certainly could have exceeded the 214,000-liter daily deficit in the aqueduct that has to be redressed by taking water from the deep wells. It seems that returning the streams to the health they possessed before the construction of the aqueduct might be a far cheaper and more straightforward solution than continuing to engineer a managed system of water provision.

GENTRIFICATION

Gentrification in this region was marked by a change in the primary vocation of the land, mostly due to the industrial cultivation of tobacco and the arrival of tourism to Barichara. The original inhabitants made way for tourists and a new, wealthier influx of people. The region has also been hit, in recent years, by the great water crisis of 2010, in which an intense summer heat wave almost completely depleted the municipality's water sources. This stoked an environmental awareness around water care (which spurred the foundation of the Senderos del Agua/Water Trails project). This wave of concern led to the creation of Bioparque Moncora, the establishment of the Green Belt, and the modification of construction to maximize water sufficiency, with an environmental force arising in Barichara that was willing to take care of the territory and to watch over the sufficiency and sustainability of its water.

Finally, an important, concrete, and symbolic event occurred during this time. Some of the cisterns or water sources that had been abandoned and in disuse during the last years, such as those in the Bosque de los Aljibes, even at the height of the water crisis, were opened again to the population. This act pointed to a possible path of resilience for the territory, where the management, care, and use of water could once again be in the hands of the communities in close connection with the care and regeneration of the local ecological composition of the territory.

RESOLVING THE IMPASSE

In July 2021, the Senderos del Agua team took the lead in cleaning the creek; a group from the community and the children went out to clean one of the most polluted areas. With live music, enthusiastic children, and collaboration from the local recycling company, they managed to remove about 20 sacks full of trash from the water. After removing that mountain of trash, they found a live tortoise underneath it all on the stream's shore. "This is the proof of living hope!" declared Alejo, one of the children.

In November 2021, a meeting was held and the research was written up in a report for the funders as a conclusion to the Senderos de Agua/ Water Trails project. After analyzing interviews and maps, the conclusion was that the children's dream of cleaning up the creek and restoring the water table as a way of addressing the water crisis—however romantic it might seem at first—was actually, given the current circumstances, the most feasible, sustainable, and economical solution for water safety in the region. By regenerating the stream as a source of clean and pure drinking water, the town could free itself from the system of controlled water supply, return to its natural relationship with the water in the area, and have ample water once again instead of the constant shortage they were facing.

"IMAGINE THE CREEK RUNNING CLEAN"

It may been ambitious but it was necessary to transform the town's overall relationship with its largest body of water and conceive of the stream as the source of life that could sustain the territory. How does a community become involved in change in a country as divided as Colombia and in a region where the water system has almost collapsed? Felipe Medina constantly emphasizes that the inner resource of belief, once activated, infuses light into a most somber past. A small seed—"imagine the creek running clean"—was enough to galvanize the whole community. Imagination was the foundational step in reading the language of water.

WHAT HAPPENED NEXT

The Senderos del Agua/Water Trails project was over. What next? Now that the children had cultivated their own vision of a water supply in Barichara, they were arguing to make this vision a reality.

In November 2021, the town held a public local crowdfunding event, Tejiendo Fondos Semilla (Weaving Seeds Fund), for community members to present their projects and propose new ideas. After the official projects had been presented, a group of four children asked the presenter for the microphone. Speaking to a staring crowd, the children said that their dream was to save the Barichara Creek and they wanted the community's support. The discourse was short, sharp, and to the point. The presenter asked Felipe to interview the children then and there, in front of everybody, to better understand what they were saying and how they could be supported. After the interview, two of the children looked directly and piercingly at Felipe and asked him to commit his support publicly. "Felipe, you are gonna help us, right?"

Together, the adults and the children needed to decide how to finally solve the current water woes of the town of Barichara. Should they go with the aqueduct and water companies and legal wrangling of the government or engage the community, clean up the creeks themselves, and get free-flowing water in the region again?

With money raised during this event, the follow-up project Pasos del Agua (Steps of Water) was begun in January 2022.[12] Pasos de Agua (Festival of Water) is a larger project with a big dream to create the first water festival in the region, an invitation by the children to elicit collective cultural creativity to celebrate the gift of the water. Through this very local festival, they seek to create a series of artistic publicity campaigns to address the value of water and the importance of the protection and restoration of the creek, including a pedagogical concert. Finally, they will host a joint conversation with the children, the community, and the local government about multiscale community-based water solutions for the future.

Cathy Holt, from Topeka, Kansas, heard about this work and moved to Barichara. She participated in a Pasos del Agua walk in December 2022[13]. She wrote about her experience:

> The state of consciousness in a community is reflected in how they care for their water. Last week, Felipe Medina, his wife Alejandra, and a few other parents took a group of 15 children, aged five to eleven, on a four-day walking journey. "Pasos de Agua" (Steps of Water) took them along the Barichara Quebrada, upstream of where the sewage goes in at Salto de Mico, or Monkey Jump. The last two days were a campout at a coffee farm near the stream, culminating in the children's public report of their dreams for the Quebrada and for the future they want to see.
>
> I was able to attend thanks to Carlos Gomez, a lover of children and big proponent of clean water, who owns the coffee farm Agua Santa (Sacred Water). He kindly offered a ride to me.
>
> While Paúl's 11-year-old son Alejo solemnly beat a drum, one younger child at a time read from some large posters they had created with adult help. Talks were given by three boys and a girl: Dante, Coco, Lorenzo, and Quetzal. Dante, perhaps 8 years old, especially got my attention with his enthusiasm and wildly funny imitations of guitar players!
>
> Felipe has such a gentle way with the children. And they were eager to tell adults about how the stream could be protected and improved.
>
> Children's requests and dreams:
>
> - Don't throw garbage in the stream, the woods, or the beach.
> - Create dry toilets.
> - Make dikes to trap the rubbish in the stream.
> - Place filters in the stream so that water comes out more pure.

- When we see anyone dumping trash into the stream, tell them "No."
- Reduce the sewage going in by directing waste into septic tanks away from the water.
- Create a campaign for people to stop using plastics and other products that contaminate the water. Publicize the issue with big posters and on TV.
- Talk with the mayor.
- Get together each month to keep developing these ideas.
- Carry out the ideas.
- Keep visiting the water.
- Let's save the stream! Water is what gives us all life. Without water we cannot live!

I asked the children how they would prepare for a meeting with the mayor, and there was some discussion among them about this. Then some kids got into the stream for a little swim.

Along came an iridescent blue morpho butterfly, color of sky, symbol of hope, transformation, and new beginnings. There is a legend that when we see this butterfly, our wishes can be granted.

The children could see beyond the filth and degradation to hope and opportunity. From its birth, the project had been a child-inspired, community-based project. In the three pilgrimages they did, the children were the main instigators and main investigators, looking for answers that could transcend the social and political biases that sometimes determine the "official" narratives of territories. It was the children who articulated most clearly the simple solution to the challenge of the water crisis: to clean up the creek and restore the natural water system. In tending to their children's dreams, the town discovered the most strategic, the most natural, the most harmonious, and the most culturally ambitious model for the revival of water in Barichara.

How often do citizens and leaders say they want to leave a better world for their children but feel powerless to make the decisions needed to do so? The community of Barichara, both adults and children, have shown what can be possible when the voices of the future are taken seriously.

Adults and children agree.
Photo courtesy of Felipe Medina.

The steps into the depths of the current water crisis in Barichara, as with our global climate crises, show us a brighter future with the children who will live in it.

POSTSCRIPT

In January and February 2024, after representation from Felipe and the children, the mayors of Barichara and five other districts held many meetings locally and also asked The Flow Partnership about accessing the wisdom of the global communities. These mayors are not only inspired to put their power behind addressing the dire water situation in their regions but have vowed to take firm steps to do so. An initial proposal has been drawn up to restore the creeks with public and private participation—a huge victory for the children in bringing about the change they want to see in the world!

This action can be directly traced back to the first walk of the Senderos del Agua/Water Trails project with the children and is a lesson for communities everywhere. Wherever we are in the world, whatever the form of the climate crisis that is facing us, this project shows us how to motivate a community, first in their imagination and then in their actions, to restore a healthy water cycle, not just for themselves but for their children and future generations.

This chapter was written in collaboration with Felipe Medina and Alejandra Maria Camargo of the organization Water Trails in Barichara, Colombia, through many interpreted and transcribed conversations. As we spoke to them, it was clear that their words, so perfectly descriptive, needed to be a part of this story.

16

The World Water Bank

To remain a vibrant, well-regulated, life-supporting blue planet, what is the real economy and the real wealth we will need? Right now, our planet is spiraling into an untenable relation with its water, which is reflected in the desertification and flooding that we are seeing everywhere. In the color insert, we see the temperature change maps from NASA over the last 50 years or so. Extreme weather events all point to an increasingly adverse relation to water. Solutions to this situation have clear parallels with the financial principles that regulate the global economy.

Consistently working with local communities who have revived their water sources and reduced droughts and floods by holding the water in their landscapes has led us to conceive of water management as the biggest banking system on the planet, bigger than all the world's currency combined. To understand the world's water bank fully, we need to go into the history of both currency and water conservation.

FINANCIAL CAPITAL AND NATURAL CAPITAL

While the reality that financial capital is dependent on natural capital is understood by most people, understanding the two systems through the lens of water lends a whole new meaning and sense of urgency to the climate change scenarios facing us.

Just as banks must have adequate cash in hand to meet the withdrawal requests of depositors on a daily basis, water banks must have adequate water on a daily basis to meet the needs of people and all other beings living on the planet. Failing to do so can adversely affect the landscape and its inhabitants, as well as diminish trust in the region's ability to support life, spurring migration.

For our economy to thrive, banking services must spread across the different sections of society, providing services to even the underdeveloped regions. In the water economy, water banking services must also be spread across the different sections of society. Little ponds scattered all over the planet for use by everyone will reduce inequalities and the impacts of climate change. Furthermore, the underdeveloped regions or rural segments of society will be able to remain in their local communities if they have access to water banking services to live their lives in situ. Migration to the cities is often the result of lack of water in the rural landscape.

Another parallel between banks and water banks is the notion of investment. Banks must have a sound investment policy to achieve the goals of profitability as well as liquidity. Water banks must invest in the future by allocating adequate area for holding water and planting trees to achieve profitability (long-term underground water recharge and a continuous water cycle) as well as liquidity (availability of surface water). In both cases, laws and regulations are not enough—the system is entirely dependent on community members carrying out these actions on a daily basis.

When currency was first invented, it was either in the form of a precious natural resource or backed by a natural resource such as cowries, livestock, salt, silver, gold, or grain. From 800 CE onward, in the western world, the English pound corresponded to a specific amount of silver (originally 324 grams) that was also equivalent to the reducible coins of 20 shillings or 240 pennies. In 1816, England introduced the gold sovereign (7.32238 grams) in its own country and in its colonies, converting to the backing of currency by gold. Gradually, France and Germany followed. In 1873 the gold standard was adopted globally when the United States followed suit.

During these periods in history, the most essential natural resource of all, water, was seen and understood beyond any doubt as the lifeblood of the world, whose management fed the establishment of civilizations and the settling of nomadic peoples into a life of prosperity. We can imagine this understanding flowing 4,000 years ago among the Sumerians as they established their civilization in the Fertile Crescent, between the Tigris and Euphrates rivers, through an elaborate irrigation system of canals, dams, and dikes to direct water to the fields. This knowledge was awake in the Egypt of 2,000 BCE, where the flooding of the Nile became the basis for an intricate cycle of seasonal activities to keep a small two-kilometer strip on either side of the river fertile. Additionally, as we have seen, elaborate irrigation systems were devised in ancient China to enrich the plains below. Throughout these periods, even though many natural disasters related to water or the lack of it, its place in the flourishing of a civilization was supreme. Everyone knew and understood that without water there was no life.

FAST FORWARD TO THE MODERN ERA

This balance between water, money, and civilization changed dramatically after World War II. The idea of global prosperity, independent of the natural world, began to spread. In 1971, the United States delinked the US dollar from the gold standard.[1] This momentous step opened the doors—especially in the 1980s with Prime Minister Margaret Thatcher in the United Kingdom and President Ronald Reagan in the United States—to total deregulation of the supply side of money and a surrender to "free market economics."[2] The banks issued money with the goal of stimulating constant growth, with currency without any underlying foundation. As a direct consequence of this, business became a frenetic driver of the modern world with nothing to slow its ever-increasing exploitation of the natural world as if it could do so endlessly. With no built-in natural rhythm to regulate its pace, the global economy went wild.

In this same period, the Green Revolution began delinking agricultural growth from the limitations of capacity in nature. An accepted doctrine—which, to a large extent, still prevails today—promoted the industrialization of agricultural practice. It was argued that fertilizers could add the necessary nutrients to the soil, artificially and infinitely, and water could be drawn from the aquifers through wells limitlessly. This Green Revolution centered around high-yielding crops, chemical fertilizers, the use of powerful irrigation technology, and wide use of chemical pesticides to control and standardize agriculture in the search of ever-increasing productivity in the fields, with scant reinvestment in the health of the land. The earth and its gifts were seen as subservient to the master of money that could be earned. Food as a controlled system of production for profit and loss, manipulated by a select few, impoverished a system from nature that had previously been free and abundant and available to all.

Vandana Shiva critiques the Green Revolution as a movement instigated in response to and against rural unrest and demand for fairer land ownership.[3] The newly introduced high-yield crop varieties were dependent on fertilizers and irrigation that depleted the soil health and emptied the aquifers. These monocrops were particularly vulnerable to pests, which necessitated the use of chemical deterrents. The increasing expenses of fertilizers and machinery borne by the farmers effectively put the land in the hands of a few rich food producers.

BLUE WATER ABSTRACTION

In monetary terms, the US federal debt has increased from $22 billion in 1922 to $31.46 trillion dollars as of May 23, 2023.[4] Governments around the world routinely increase their colossal figures of financial debt; it seems that running a financial deficit has become standard economic policy. Delinking financial systems from the free gifts of the natural world has brought the world into the imbalance and chaos we see today, including the income inequalities and climate inequalities (which are inextricably linked) driving this chaos.

Meanwhile, we see estimated global groundwater depletion at around 730 cubic kilometers per year[5] because of "non-local or non-renewable blue water abstraction."[6] Traditional water methods that would balance the surface water (of rivers, lakes, and reservoirs) while maintaining the aquifer levels have been replaced with using groundwater without replenishment. Between 16 and 33 percent of agricultural water use is nonrenewable (400–800 cubic kilometers of annual water loss).[7] The sum of water depletion in the United States alone during 1900–2008 totals approximately 1,000 cubic kilometers.[8] In the same period, 35 percent of wetlands worldwide were drained.[9]

In a surreal twist, as the fresh water on the planet runs out, water is now being traded as a commodity on the stock markets. We are running on fear, chasing empty wealth, and accelerating a future where there is no water to drink and no food to eat while total annihilation is staring us in the face as a consequence.[10]

Another example of this mindset is the sale and trade of carbon credits, where the ecological well-being of the world has become a competition to exchange polluting industrial practices for token green projects purportedly happening in compensation. In many cases, billions of dollars' worth of carbon credits from the biggest certifiers have been found to be worthless.[11]

If we are going to reduce the impacts of climate change and restore the planetary climate system to some sort of balance, water, money, and civilization need to be united coherently again. This is a nonnegotiable reality. It's not philosophical musing but a clear and inviolable reality that these cycles of nature have to interlink once again to restabilize the climate and reduce income inequalities, migration, poverty, crime, desertification, extreme climate and weather events, displacement, regional wars—the list goes on and on.

THE WATER BANKS OF THE WORLD

Whether we acknowledge it or not, nature is already asserting its authority in determining our financial futures. Floods, drought, and forced migration emergencies are already beginning to compromise the capacity of human systems to control the environment for its own benefit. The planet is already in command of its future, regardless of the tokens we are currently using to measure human wealth.

A country becomes rich by managing its water resources well. Local water banks, built and maintained by local communities, are the holders of their wealth, enabling a sustainable and prosperous rural life even in arid and flood-prone lands.

One of the aims of the Green Revolution was to modernize traditional farming methods in India in order to make the country self-sufficient in grain production. This required the removal of vast quantities of water from the aquifers (without systems in place to replenish these same aquifers), leading to a crisis of water debt akin to the financial debt of countries. Where traditional practices had worked within natural cycles, the Green Revolution was both financially and ecologically unsustainable in the long term, requiring debts of money and water to drive that intense agricultural system. The NASA GRACE satellite reported that "More than 109 cubic km (26 cubic miles) of groundwater disappeared between 2002 and 2008"[12] in Northern India—water that was taken from the aquifer through wells and not replaced. (See map in photo insert showing groundwater depletion.)

Rectifying the water balance requires a change in attitude. As we have seen, Rajendra Singh introduced Rajasthan to holding water collectively, with community engagement, in huge ponds strategically placed in the landscape to collect the monsoon rain and recharge the aquifers.[13] The ponds were situated in geological areas with vertical fractures leading through the rock down to the aquifer to maximize water percolation. After building many ponds with access to the underwater layer, the aquifer systematically recharged.

In the same region, The Flow Partnership started a project in collaboration with Rajendra Singh to address the replenishment of the Saseri water bank, an example of how investing in the water cycle has benefits for climate, agriculture, biodiversity, water security, and more. When we were implementing this project, we wondered how we could amplify and replicate such small yet achievable projects to make a significant difference at a catchment scale. On the financial balance sheet of the project, money was provided for labor and the engineering works that fulfilled the technical work. However, what was unaccounted for was the tremendous unforeseen wealth that emerged when the water came back. Some of this wealth manifested in increased productivity, from the convenience of having water nearby as opposed to walking miles for it, as well as children being able to go to school, better health, and provision of water for animals. We saw the landscape restored and returned to abundant life, a thriving ecosystem expressing itself with the return of birds, bees, and biodiversity. Rekindling life and reactivating the carbon and water cycle can be replicated all over the planet. The Saseri water bank, described below, is a model that can be replicated all over the world.

THE SASERI WATER BANK BRANCH

Driving along the dusty roads to Saseri in the Karauli district of Rajasthan, in India, we were struck by the lack of plastic litter on the roadside. We realized quickly that this was because there are no shops in the area to buy anything. There is no point in having any shops where people have no disposable income for little pleasures like tetra-packed soft drinks, a bag of chocolates, cookies, etc. The people here eke a living from the crops they are able to grow and live only on that. In this semi-arid climate, with summer temperatures which can sometimes reach 50 degrees Celsius and above, their existence is marginal. Many leave and migrate to the cities in search of the basics of life.

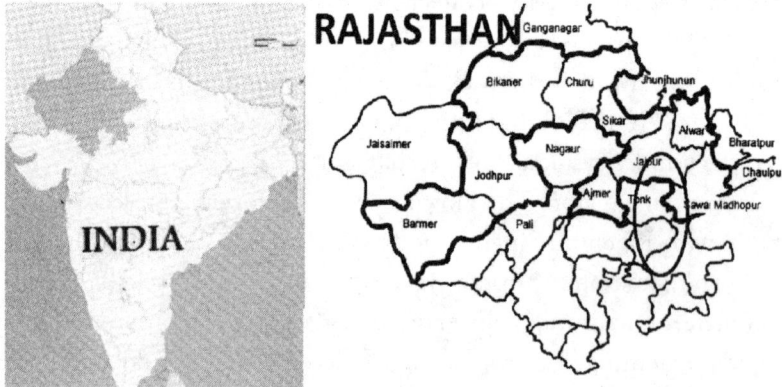

The location of Saseri.

The world has often progressed by capitalizing on the land from which simple community life has been displaced. But here in Saseri, there is nothing to buy or make capital from. Such places do not figure in national financial calculations when banks are deciding where to open a branch.

And yet, bringing health and well-being into the area is still possible.

For small marginal farmers, water held in the ground has a direct impact on the amount of food they can grow. Only when there is water will they be able to grow enough food for themselves and trade the surplus to get some money for life's basic necessities, such as sending their children to school and living in sturdy stone houses instead of flimsy mud huts.

By holding water in the ground, which is their local water bank, the farmer also helps balance the natural cycles, creating human wealth that is inclusive of natural wealth. Instead of seeing the climate crisis as separate from our economic performance, the World Water Bank is a process of returning a natural wealth that we can collectively share on a global scale.

FOUNDING A LOCAL WATER BANK

The following drone image (taken April 15, 2019) shows the state of the land before the work began.

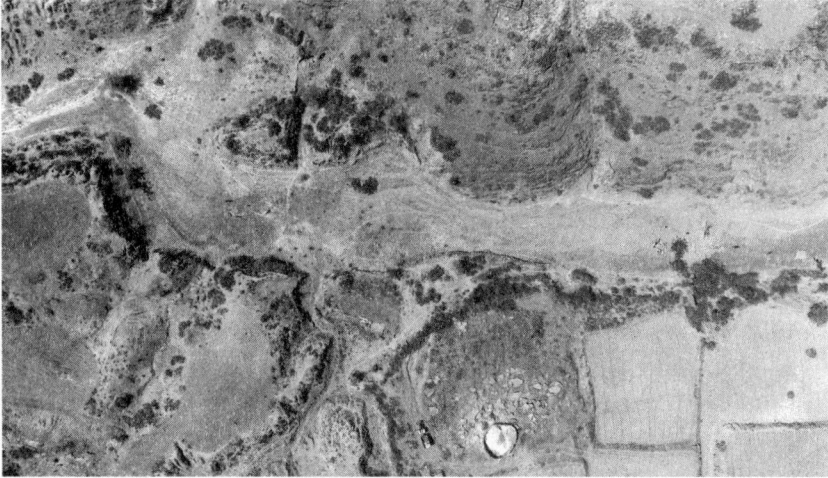

Drone image of the site with the group ceremony initiating the structure, right. Photo courtesy of The Flow Partnership.

The community takes the first step by deciding what is of real worth to them. The commodities of the cities are worthless in rural areas and, in any case, no one could afford them. Here, no water equals no life. The future they want to realize is in their own hands. They create their water bank in anticipation of the monsoon rains to hold not just water but hopes of a decent life.

Soil is dug out from the depression at the center of the picture and moved to form a mud bank on the right, to create a natural container to hold the water that will fall from the sky.

The community owns the process of creating their water bank. There is no international financial authority or business investing millions to bring their bank's branch here. The water bank is made by their own hard work, the fruits of which will accrue locally, giving them the opportunity to multiply that wealth.

The water bank under construction: thousands of tractor loads of earth, dug out from the ground to hold the water, go to building the new bank. Photo courtesy of The Flow Partnership.

The drone image in the photo insert ("The bank seals the water holding area"), taken on April 30, 2019, shows how busy the villagers have been over these two weeks, moving hundreds of cubic meters of earth to create their water bank.

The other drone image in the photo insert ("The monsoon rain fills up the structure"), taken on August 30, 2019, after the monsoon rains of the same year, shows the complete and full water bank. In just four months, they built a water bank that can hold 10,000 cubic meters of water, enough for several years of use by the villagers. As the rain falls and fills the water bank, the water in it immediately enables 27 hectares of new farmland to be seeded with wheat, ensuring a newfound food security in the village.

The previously dead, semiarid landscape transforms into a lush tapestry of green, home to many birds and animals, where life has a chance to take foothold again. (An illustration of this can be seen in the photo insert: "A lush wheat crop.") If you lose this natural measure of wealth, then you eventually lose the trade it will enable on the stock markets of the world. It is as simple as that.

HOW DOES ONE CALCULATE THE PROSPERITY AND LIFE THAT THIS WATER BANK ENABLES?

Can one even calculate it in digital money or in gold coins? Sure, you can calculate the worth of the pukka house they are now able to build, or give a value to their newly green land as posited by a property developer. But you cannot put a monetary figure on the true value of a life being lived in health, well-being, and joy or the impact on the lives of the children who can now go to school and become literate, contributing to society as they grow up; or the women who now have time to sit with friends under the communal tree in the evening after all the work is done and know peace; or on the regenerated landscape that enables healthy life and contributes to the local and planetary water cycle. This is the primary wealth that is worth realizing and storing for future generations in the local water banks!

SHAREHOLDERS MEETING

Many wealthy people today would question whether the money they keep in their bank is worth anything at all. With increasing natural crises, the volatility of commodity prices, suspended stocks and falling shares, what is that money actually worth?

This water bank shareholders meeting described below has a new, prosperous feeling to it.

A shareholders and water-holders meeting November 2019. Photo courtesy of The Flow Partnership.

The local community of farmers and villagers and the international community (from Arup Engineering, the WaterUP team of The Flow Partnership, and project leaders of the NGO Tarun Bharat Sangh)—all investors in the water bank at Saseri—held the first shareholders meeting of the new community water bank in 2019. They are very satisfied with their investment in this bank and can see the visible transformation their investment has enabled. Is this mutual feeling of sharing in a renewal of land and life the real, true wealth?

The Saseri rural economy is in circular motion again. The turnaround has taken place. The community is now the one holding the real wealth. Currency has also been transformed by such a process and the economy has started flowing again.

WITHDRAWING FROM THE BANK (REAPING RETURNS)

The first crop for the farmer and family on the newly watered lands. Photo courtesy of TBS.

The Saseri village water bank structure, with a realized capacity of 10,000 cubic meters of water, creates wealth in terms of an improved landscape, resilient climate, rural renewal, livelihoods, and credit in the money bank for the villagers. From a local, small-scale water bank, the enhanced local prosperity will be exchangeable into globally recognized wealth.

The villagers are proud of their work and own a share in the water bank they have created. This is a different type of ownership from money in individual accounts in a bank. Having part ownership of the local water bank means having a share in the transformation of the environment from arid to fertile green and a prosperous village and life.

The effects are profound:

- *Johads* (water ponds) provide local communities with access to reliable, clean water for drinking, washing, and irrigating.

- Improved water access greatly improves the health and well-being of local people from a nutritional perspective and leads to socioeconomic benefits such as improved health and more accessible education.

- Twenty-seven new hectares of farmland for growing wheat gives farmers a sustainable and comfortable livelihood. They can go to the market and sell their wheat, enabling them to build new, solid stone houses for families in the village.

- Harvests are more successful with greater yields. Some farmers were previously only getting one harvest a year, if that. Now they can reap as many as three harvests annually without impoverishing the soil.

- Villagers can sustainably keep and graze livestock and stock fish, which has further enhanced their livelihoods.

- The aquifer recharges and the water imbalance is corrected.

- The Siberian stork has been seen again in the area on its migratory passage.

- Collectively local incomes in the area increased by approximately £10,000, benefiting 1,200 people. Nine hundred new domestic animals are now reared for milk and other products.

- When girl children no longer had to collect water with their mothers, school attendance increased.

- Youth who would have joined local bandits when the water vanished are now gainfully employed selling milk.

- With the additional time, the women were able to take advantage of a government scheme to buy natural gas cookers for their homes so they would no longer have to forage for wood for fuel.

- Crucially, during COVID-19, 120 migrants who had to return to these villages have more food and water to recover and live than those who returned to neighboring water-scarce villages.

RETURNING TO AN OLD BANKING SYSTEM

What if a time comes when the pace of extreme natural crises increases so much that financial indicators such as GDP, stock markets, and corporate profitability can no longer pretend to hold the wealth of the world?

Can the wealth generated by these local water bank branches, which provide regional stability and local resilience to the landscape, become the measure for future growth?

Financial institutions are holding paper wealth—now digital and even less material—as if that was the only wealth the world needed to look after. That money is as invisible as its power while its impact vanishes with a depreciating climate situation. No amount of slick smoke and mirrors declaring otherwise can hide the fact that our planet and civilization are in deep natural trouble.

Are those multitrillion dollars hidden in vaults or digitized on computers going to be kept hidden and dead or can they be made active to help resolve climate change, natural disasters, inequality, hunger, and poverty?

The multiple crises today are showing us that the real source of wealth is not commodities, coins, or currency but community action, nature, water, and a healthy landscape.

Annually, a global freshwater deficit of 730 cubic kilometers of water is flushed and wasted down drains, treating it like a single-use commodity and removing its capacity to infiltrate the ground and restore the local small water cycle. The local water bank proudly makes local communities the keepers of this genuine wealth. Our challenge on a global scale is no different.

INVESTING IN THE FUTURE

As financial banks around the world fund their virtual wealth through cycles of increasing debt, so the water banks around the world are being re-dug to turn back the debt from the water cycle into a positive natural wealth for people and planet and peace on the land. In the future, the creation of money should be on the authority of nature, supporting the natural wealth of a healthy water cycle and planet.

The world can pay back its trillion dollars of national debt and solve the water imbalance of 730 cubic kilometers of global water run-off (730,000,000,000 cubic meters) by building 73 million Saseri-type structures holding 10,000 cubic meters each! This is not a fanciful notion. It is the very action required to ensure our planetary survival.

Each region on the planet can rebuild their water banks according to their unique landscape and climate patterns. A prime example of this is Water School Africa,[14] which is working to establish local water banks with communities in the Sahel and across the continent of Africa. These water banks are even more localized: each field is equipped with a relevant-sized water bank account to hold the runoff from heavy rains where it falls. The rainwater collects in each farmer's own water bank account and allows crops to grow, generating financial wealth for these farmers.

A small portion of our financial wealth can be held in these water banks with the return of regenerating a healthy planet. Our high-dividend payments come not from a shaky financial system that makes only a handful of people rich at the expense of all others, but from a vibrant healthy planet that enriches all those who live on it.

The healthy national water banks of each country will revive rivers in desert areas and stop the buildup of floodwaters before they reach our homes. They minimize erosion, recharge aquifers, regulate localized weather patterns, and sequester carbon. They also create visible local, national, and global wealth for all living beings on the planet. (See the photo "The stages of the process" in the photo insert.)

A POTENTIAL PATHWAY

The Red Cross became the seminal contribution of the 19th century to the world. A similar global organization, effective from its highest levels to the grassroots, to hold the water on our planet is needed today. Philosopher and activist Lorenzo Marsili, in an article in *The Guardian*, proposes the idea of a Green Cross, laying out what could happen if we create "a planetary civil protection force with rapid-reaction capacities and clear operational expertise guaranteed by regular joint training and funding. When wildfires strike in the north of Africa, nearby teams from Europe, the Middle East, and sub-Saharan Africa could jump in knowing exactly what to do. When flooding wreaks havoc in south-east Asia, Chinese, Indian and Australian corps would know how best to coordinate relief."[15]

Of course, there are plenty of ideas around, and they cannot all work. But underneath the current crisis is a vital choice. Are we unified behind a return to harmony in human ways of being on the earth, or are we into the thrill of betting how far the fragmentation of the earth's resources for monetary profit for the few can take us?

Whoever you are, wherever you are, no matter if you live rurally or in a city, you can do something about bringing water security back to our planet. Take responsibility. Join your community. Hold the water in the ground. Research your water catchment and give it what it needs to thrive; if you can't do any of that, then support a community that can. Earth is our only real shareholder[16] and local water banks the only meaningful route to becoming truly wealthy.

CONCLUSION

What does a movement that could rebalance the landscape and climate for both the city and the village look like?

When we began this work of holding water with communities, there was still an assumption that centralized, hard engineering solutions would take care of the management of droughts, floods, and other such problems. The development of engineering techniques gave the illusion of keeping the landscape healthy by controlling it. The idea that humankind's industrial patterns of behavior were capable of undermining climate stability beyond all previous limits was still a theoretical discussion and didn't seem worthy of any airtime.

When we looked more closely, what we saw instead was the precariousness of environmental health in spite of the application of all those hard engineering solutions. We realized that we couldn't buy the assumption that the situation was under control, so we started digging around to see what really was happening in the United Kingdom, where we live. A few pilot solutions mimicking and harnessing the techniques of nature to keep landscapes healthy (called nature-based solutions) were being tried, but generally these were seen as minor, unproven approaches with no evidence to support them. Hard engineering—large, artificial structures that try to keep natural processes such as excess water or lack of water under control—was widely relied on, practiced quite separately from the water that flowed in the landscape. The wisdom of soft engineering, such as small, traditional practices around holding rainfall in the land, was treated as incidental or folklore. Solutions drawing from traditional knowledge of nature were being implemented on the scale of entire rivers and river catchments in places like the dry state of

Rajasthan, India. But these were happening completely out of public view and without a global discussion around community-driven water retention.

After a chance meeting with Rajendra Singh, he invited us to visit him and his communities in Rajasthan to see firsthand their work holding the monsoon rains in a pond structure called a *johad*. The villagers explained to us that the structure is shaped like a half-moon to collect the water just where the rainwater runoff from the hills aggregates. Surface water becomes instantly available, and the seepage from that water goes down to the underground aquifer through vertical fractures below, which replenishes once the *johad* starts filling up. Once the rains come and the *johads* begin to fill up, the countryside becomes quenched; the fields are like a cell full of renewed energy and bursting with green crops; the air is full of the songs of birds and insects; and there is a full-scale communication of life all around. Was this Rajasthan? We had heard that it was a desert state with little plant life, and instead we felt as if we were in lush, green England (though the colors of the land and the camels pulling carts belied that, of course).

Back in the United Kingdom we made an even more surprising discovery. In Belford, they had built structures similar to *johads* for holding the water—very like the ones we had seen in Rajasthan—but this time, the intention was to stop the vulnerable villages there from *flooding*. The man-made, pondlike structure in the landscape of Belford was used to temporarily slow the flow and sink the stormwater into the ground so it did not congregate somewhere else, like a low bridge over a stream, and flood the houses and the roads.

It dawned on us that, of course, floods and droughts are two sides of the same coin—that of managing (or mismanaging) the flow of water in the landscape. If we know this, then why do we allow excess rainwater from storms to gather momentum into floods or let the landscape dry up so much that even the underground water vanishes and there is a drought? Why were communities and landscapes around the world becoming desertified and the dry regions of the world going thirsty when there was a simple, natural process to hold the rainwater as it fell from the sky?

We contacted hydrologists Paul Quinn and Mark Wilkinson from the James Hutton Institute in Aberdeen, Scotland, at once and gave a presentation for them and their scientists, demonstrating, to their amazement, how their success at Belford was mirrored by thousands of completed structures and seven revived rivers in Rajasthan, India. A few months later we visited Michal Kravčík in Slovakia, who has also, with his local communities, constructed ponds, leaky dams, and swales nationwide, to stop flooding in over 800 villages there. In 2014, we invited Rajendra Singh and Michal Kravčík to present their work to communities, scientists, and policy makers in the United Kingdom. We were also invited to meet with King Charles (then Prince Charles), who had already visited Rajendra Singh's communities in Rajasthan and had been quite impressed by their successful, village-level techniques for holding water.

Despite climate change being in the news now almost every day, there was no global conversation about water with community experts who could restore water cycles in the landscape. The Flow Partnership became a network and platform for the expertise of community "soft engineering specialists" to finally be heard. Our ambitious aim was to bring together the structural solutions of engineering with community practices of holding water. We were locating the alphabets of the language of water in our discoveries and stories from working with global and local communities—all of which have now coalesced into this book.

Every day, news outlets give us a large dose of environmental crisis. The stories are presented with a mixture of scientific neutrality and intensifying on-the-ground reports of climate disasters, passively presented as if these are unavoidable consequences of climate change.[1] There is much denial, incredulity, outright dismissal, and even obfuscation of the techniques that communities can apply to address our changing climate. Yet the only division around the heating of the planet to dangerous levels seems to be in the scientific community—not in local communities living on the ground in countries all over the world.

Since founding the Flow Partnership, it has become very clear that the stories of practitioners on the ground still remain largely untold. Sustainability conferences, policy reports, research projects, and academic authority have left little room for the unfiltered voices of those who are conducting the actual work. The language of water demands a school to resource its speakers, to communicate its stories, to spread its simple message. There needs to be a systematic platform where these natural solutions to the problem of too little or too much water in the landscape can be shared, disseminated, and accessed by local people in simple, easy-to-understand, practical language. Part of the reason that communities were not holding and collecting water in the landscape from which their lives could flow was that the knowledge of how to do that effectively was not always available.

One of the Flow Partnership's powerful alliances is with global civil engineering company ARUP, which came about through a chance meeting on a train with one of their geologists, David Hetherington, who saw the potential of hard engineering and soft engineering coming together to benefit local water cycles. ARUP joined forces with us in 2018, along with the local community in Karauli, Rajasthan, India, to implement a shared water-retention project called Water Up. The aim of that project was to create successful traditional water-holding structures with the local people of that area and to film their methods and practices so they could be disseminated in simple, language-neutral tools that communities across the world could access, to implement water-holding structures in their own regions.

After the *johad* had been built, ARUP engineers, project leaders, and local community members met on the bank of the completed structure and mutually appreciated the quality of transformation that holding the water from the monsoon rains had brought about. Where the engineers proceeded from a closed design, backed up by calculation, the villagers worked from an open relationship to the landscape, with the old knowledge of how to hold the water using traditional methods. As long as both were done with quality, there was total complementarity between the two.

By bridging the civil engineering knowledge of ARUP with the Rajasthan communities' traditional, natural solutions to water scarcity in India, we were also able to understand the gap between these two approaches. This intersection of engineering knowledge and community wisdom around the language of water presents profound implications for dealing with global warming by spreading this work with local communities on a global scale.

The lessons from Water Up led to the establishment of the global Water Schools as platforms (both online and on the ground in local communities), which allow communities to share their water practices and wisdom with each other. They go a step further through innovative funding mechanisms that support communities in implementing water retention actions in their area to restore their water cycle.

Some challenges they address:

Understanding the Role of the Water Cycle

Local action on the water cycle can have a profound impact on cooling the planet. Everyone is talking about the dire circumstances we will be in if we do not reverse global warming, but few people are talking about how local water-holding work can be the key to this change. In a project the Flow Partnership contributed to in Kanpur, India, in 2019, we researched the impact of community practices on surface temperature and crop health. The surface temperature as shown by satellite data was five or six degrees Celsius cooler in areas where water from the canal continually evaporated from the leaves of rice plants compared to areas where the canal could not reach. Explaining this requires knowledge of the water cycle: how energy from sunlight is absorbed when water turns to vapor and dramatically cools the surface temperature. The whole foundation of the water cycle, from trees to climate to soil, is geared toward life having access to drink what it needs. We can change this world by cutting down the forests or letting the canal silt up, or the soil dry out, if we are happy to turn a cool, vibrant planet into a scorching, semiarid desert as we seem to be doing today. But the satellite data showed how traditional methods of tending to the plants and the forests, through the provision of running water, can influence the local temperature and plant life and can be extrapolated to regain planetary balance.

Understanding the Economics of Water

The "water bank" is an understanding of the economics of water in the landscape. A pond, a natural or dug depression in the ground, can hold thousands of cubic meters of water falling as rain. The walls of the pond are an earthen bank that keeps the water from flowing out. This body of water can be seen as a credit account within a water bank. The community then has this water credit to grow their crops, feed their animals, and profit from seasonal sales of vegetables and milk. We can imagine a whole financial system where the water bank owned by the community replaces the financial investment by the monetary banks of the world. In one, the wealth and profits are widely shared, keeping the world alive and healthy, while in the other, as is increasingly evident, a dead currency benefits only a few and is unsustainable for the earth.

Action and Inaction

Most people understand the dire consequences of climate change, but few understand that the key to reversing it is in local water-holding work. The people who actually take the actions necessary to restore the water cycle and reverse climate change are often given second place and kindly patted on the back if they say something outside the context of science. The world forgets that no amount of scientific papers, evidence, numbers, and analysis can take the place of digging the pond and holding the water or growing the tree that keeps the water cycle healthy and cools the planet. It is the action taken by people and local communities that will restore the stability of the climate and cool our planet.

Funding

The communities who do the work efficiently and for little money are not served by investment channels. Funding usually supports large projects that garner huge headlines but achieve little impact on the ground for the amount invested. Yet people who invest wonder why there is no positive difference in the climate when so much money has been continuously given to climate change projects! Unless we recognize and resource the millions of local communities tending to the land and restoring climate balance, we will just limp along from crisis to crisis.

There are many books approaching water from different perspectives—from the policy perspective, from the scientific perspective, from the financial perspective, from the perspective of alleviating poverty, and from the perspective of large NGO water projects. The language of water unites all of these. This book approaches landscape rejuvenation from the local community and people's perspective, and their agency, to speak the language of water and deal with climate change at our level.

With engineering solutions that are drawn up as separate from nature, humankind has created new languages of structural mechanics and digital design. Nature is gradually being relegated to a passive receptacle of already structured solutions, leaving scant room for any imagination or self-responsibility or agency. No wonder we feel powerless to do anything about the climate challenges facing us today!

How can we come back to basics and first understand our part in the water cycle, whether we live in a city, a village, or somewhere between? What strategies can city dwellers employ at their own individual level to connect to their local water cycle even if they live in a small flat in a high-rise apartment without access to land where they can hold rainwater or plant trees?

With climate change upon us, the scientist and the farmer need to work together to understand the multiple natural factors that are out of balance and imagine together the solutions for a planet that is heating up and destabilizing to dangerous levels.

In rural areas, communities on the ground observe nature day in, day out, as only they can, with the land and surroundings serving as their laboratory. Their experience of growing food and trees gives them unique insight into the signs of impending floods and droughts and the methods to minimize their occurrence. This is a lifelong experiential PhD—not written up in academic papers but in the language of nature and life in a healthy landscape. The proof is in the food we see on our shelves.

In urban areas, scientists are exploring, researching, and analyzing what will return the planet to health through complex data models. They go to these same farmers and communities to collect their data to understand the processes of nature. It is to a large extent forgotten that the local methods of water retention practiced by the farmers and the communities are the origin of our scientific understanding. Scientists often discover that communities understand a natural phenomenon or process very clearly but have no voice or platform to express their wisdom and be heard.

We have observed again and again educated people, even with the best of intentions, trusting that the ultimate solutions rest in science. When the "illiterate" villagers start to speak, they are often misunderstood or asked to give evidence of their observations. When people begin quoting a plethora of papers and analyses, overriding what the villagers are saying, the villagers fall silent out of a mistaken sense of inferiority and shame for not knowing the language the scientists are using. Yet science is only interpreting the language of water, something the villagers understand fluently!

As we have seen in the stories in this book, we need both—we need the speaker of the language of water and the interpreter with a big-picture analysis. If we are to deal with climate change and bring the planet back to health, we need those living on the land to know what to do and actually do it at a local level. From that local action, observation, and wisdom, we bring in the scientists with their skills of interpreting and analyzing what needs to happen at a global level.

This book amplifies the voices of local communities, bringing the inevitable understanding that small, local actions are crucial for the big changes that benefit us all. These small, local actions can be done at scale, eventually weaving a vast web of interlinked regeneration across the planet. There aren't enough scientists to take that level of action, but there are enough communities, willing and able to restore water security for our planet.

And then we arrive at our own agency in bringing about change. First, we have to stop and assess how the water cycle works in our local environment. Researching where our water comes from is at the tip of our fingers. We have to do our homework to understand it well. Trust that it isn't too complicated to understand; after all, we learned it in our very earliest years at school.

Then we have to improve upon what we see and learn, kickstart our imagination by really experiencing the weather in our place of dwelling and observing the pattern characterizing it. No matter how basic our questions seem, they will lead to a new depth of understanding and a belief that our imagination can bring the reality we want into being.

QUESTIONS

Where does the water in my tap come from? What happens to the rain that falls during a storm? Where does it go after it goes down the drain? What am I doing that is helpful to the water on the planet? Simple questions—almost too simple. Without knowing the answer to them, the steps we can take are shrouded in unknown externalities, leaving us bereft of our own agency to do something about them.

Exercise 1

Observe your local weather each day for three days. After that, see if you can recognize a pattern or predict what might happen. What might the weather be like in the next three days? This gives us an experiential understanding of how our climate works. We can have this experience of the weather anywhere, any place, any time. Be curious.

Exercise 2

When you take a shower, wash your vegetables, or throw away that half-drunk glass of water, imagine the journey of the water you have used. Where does it go from your bathroom drain or kitchen sink?

QUESTIONS

Does that water you just flushed away go into the ground and rise to the sky and enter the water cycle from the land outside your house? Does it enter the sewer system and get transported to the waste treatment plant and then out to sea? Perhaps both?

Exercise 3

Go outside and actually see where the water is going—where does the drain from your apartment or dwelling lead to? Follow it through your city until you know which body of water it ends up in and what is happening to those gallons of water that left your apartment just a few minutes ago.

When we know what is happening to that water we have just used, flushed away, or drunk, following its route from the apartment and right through the city until it reaches the body of water it ends up in, suddenly that water creeps into our awareness as something powerful—and we start feeling more responsible for it.

QUESTIONS

Do all drains lead to the ocean? What does that water you have just used do to the local water cycle of your city or region? Do you live in a city where some areas feel immensely hot in summer, with vast amounts of heat rising up from the pavement? Is the rainfall pattern changing around you, with too much rain all at once or no rain for long periods of time? Why does it seem to be getting hotter in the summers? Do you live in a city where rainwater is being drained out to the sea? Why isn't it being stored in the ground underneath the pavements and concrete? Is all this linked to the water you have just used? How?

Exercise 4

Imagine the water you have just used cycling through its various stages and coming back to earth as the rain that you can see from your window. Whether you are looking out from the 54th floor onto a cityscape or into a rural landscape, you have been a fundamental part of what enabled that rain to fall.

QUESTIONS

What if there were more trees in our cities? What would happen if there were rain gardens holding the water? What if water from roofs was collected for a local supply to supplement the water system? What if there were urban gardens on all the roofs, like there are in Cuba?[1] Mickey Ellinger in *Reimagine* magazine gives us the numbers: "Havana, with 2.5 million people, has more than 200 gardens, plus thousands of backyards and rooftops where people grow leaf vegetables, tomatoes, herbs, and even wine grapes."[2] What if we were to grow what we could, no matter which city or town in the world we live in, even on our terraces and windowsills if that is all the space we have?

Exercise 5

Plenty of free advice from successful indoor microgardeners for any city in the world is splashed across the internet. Go deeper into these questions and look into building or joining a community garden. We can either find where they are located and join them or, if there aren't any, we can start our own! These small, revolutionary acts on our part will set water cycles in motion to become healthy again.

QUESTIONS

Finally, we can ask what impacts of water-retention measures upstream in rural areas affect the security of downstream environments such as towns and cities? When upstream forests are planted, is the city better protected? Should the soil on farms be made more porous to the rains? Could there be ponds to slow the flow of water into the downstream towns? What is the next level of the water cycle we can affect within our own sphere of action and influence?

Exercise 6

Wherever there is a little ground in the city, we might be able to build a little drain that sends the water into the ground instead of being carried away to the sea. If you live in a bungalow, can you trap all that rainwater in the verges and grass around the house and help it to go into the ground? If you live in an apartment, can you use water wisely? Can you engage with your apartment community to look at water-holding measures that can be initiated there?

When we were working in Slovakia, we saw simple pipes being inserted into the grassy patches outside apartment buildings to allow the rainwater to go into the ground instead of being taken away by a stormwater drain. This kind of action begins with a conversation with our residential community. The situation is now critical enough to engage with our own localities and administrations to start holding the water, planting more trees in the neighborhood, and even simply creating more awareness around using water wisely.

Choosing not to take these actions is a luxury we no longer have. And don't worry about instant results. The point is to initiate the action and trust that it will bring about the results needed. We can take our example from the film *Return of the River*,[3] in which a dedicated intergenerational group of citizens kept working with their local politicians and administration to eventually stop, decommission, and dismantle two working dams so their river could run free again. The two main political parties, Democrats and Republicans, and three generations of communities in the state of Washington worked doggedly until that aim was reached. Never underestimate the power of persistent community engagement, even in the face of hopelessness.

Reality is powered by our imagination. It is with our imagination that we first see what is possible.

We have to imagine our role in making the change we want to see on the planet. We have to imagine the planet green again. We have to imagine our water cycles functioning again. We have to imagine what we are reaching out toward before we can take the steps to reach it. The scale of global warming and our inability to do anything at the macro level sometimes overwhelms us so much that imagination often scuttles off to hide underground, leaving us instead with hopelessness, despair, and impotence.

If all we have is our imagination, voice, and actions—let us not be afraid to use them!

We too can hold the water right here, in our houses, our apartment blocks, our neighborhoods, our towns, and our cities. Once we can picture the relationship of our own local water environment to the larger context, we can collaborate with others in imagining how to enhance the whole catchment or ecosystem. The language of water that expresses itself in each drop, each pond, each community, each revived stream is also spoken in the flow of different actions that can contribute to a healthy catchment. The language of water naturally combines how one person sees their actions within a city environment with the perspective of the farmer deciding how to manage their hectares of land upstream. When speaking the language of water, our own imagination is expanded by sharing it with our neighbor. If our food comes from the rural farmer whose crops are nourished by rain that comes directly from our actions in the city to keep the water cycle turning, then there is no doubt that each of our individual actions are connected to the water cycle that keeps us alive and resilient in the countryside and the city.

An apparently arid area without signs of life is not a dead region. A water cycle that has been severed by deforestation, soil drying out, and rain runoff *can* be restored. The preservation of the seed in the dry soil carries within it the whole cycle of life.

Urban resilience and rural resilience are mutually dependent. Towns and cities need the surrounding countryside to hold the rain in soils, reservoirs, and as groundwater upstream so that it is there for downstream accessibility. However, if the urban water cycle is degraded by draining city water into the sea, or vast heat islands push the clouds away and create their own destructive weather patterns, then urban realities begin to demand more of the rural hinterlands, adversely impacting rural resilience.

The language of water is now an endangered language, but we can learn to speak it fluently again. Our biggest vehicle for scaling up these local solutions is ourselves, those of us who form the millions of local communities in every country of the world, whether we live in cities or villages. The solution to climate breakdown and refilling the water banks of the world lies in the actions of all of us in our local communities working together.

Earth really is our only shareholder;[5] the investments producing real wealth are secure water banks on a healthy planet.

MINNI JAIN AND PHILIP FRANSES
THE FLOW PARTNERSHIP

NOTES

Introduction

1. Dr T. V. Ramachandra, "Bengaluru can become worse than Cape Town if mismanagement of water continues," *The India Express*, April 4, 2024, https://indianexpress.com/article/cities/bangalore/bengaluru-water-crisis -cape-town-mismanagement-t-v-ramachandra-9250290/
2. Jonathan Watts and Tural Ahmedzade, March 16, 2024, "Scientists divided over whether record heat is acceleration of climate crisis," https://www .theguardian.com/science/2024/mar/16/scientists-divided-record-heat -acceleration-climate-crisis.
3. Barnes, M, Zhang, Q, Robeson, S, Young L, Burakowski E, Oishi A, Stoy, P, Katul, G, Novick, K; (2024) Century of Reforestation Reduced Anthropogenic Warming in the Eastern United States Earth's Future; 13 February Vol 12 Issue 2.

Chapter 1

1. "The Water Man of India Receives Stockholm Water Prize," 2015, Stockholm International Water Institute (SIWI), https://siwi.org/stockholm -water-prize/laureates/2015-rajendra-singh.
2. Water Up project, https://www.arup.com/projects/waterup.
3. Water Schools, www.waterschools.org.
4. Church of England Net Zero Carbon Routemap, https://www. churchofengland.org/about/environment-and-climate-change/net-zero -carbon-routemap.
5. Apricot Centre Huxhams Cross Farm, https://www.apricotcentre.co.uk /huxhams-farm.
6. The Flow Partnership, "Water for All," https://www.youtube.com /watch?v=JKXGfyR5_SY.
7. B. Cronin, 2016, "Keeping Pickering Flood Free," *New Civil Engineer*,

accessed February 29, 2023, https://www.newcivilengineer.com/archive
/keeping-pickering-flood-free-29-02- 2016/.

8. "Natural Flood Management Programme: evaluation report," Environment
Agency, https://www.gov.uk/government/publications/natural-flood
-management-programme-evaluation-report/natural-flood-management
-programme-evaluation-report.

Chapter 2

1. A. Mishra, *The Radiant Raindrops of Rajasthan*, original title *Rajasthan Ki Rajat
Boonden*, translated by Maya Jani of the Gandhi Peace Foundation, accessed
October 5, 2023, https://www.arvindguptatoys.com/arvindgupta/anupam.
pdf; A. Mishra, *The Ponds Are Still Relevant*, original title *Aaj Bhi Khare Hain
Talaab*, Gandhi Peace Foundation, accessed October 5, 2023, https://www.
indiawaterportal.org/articles/aaj-bhi-khare-hain-talaab.

2. A selection of Anupam Mishra's works is freely available at https://www.
indiawaterportal.org/articles/aaj-bhi-khare-hain-talaab.

3. Montaut preface in Mishra, *The Radiant Raindrops of Rajasthan*, 3.

4. "Anupam Mishra: The Man Who Dreamed of Self-Sufficient India," Down
to Earth, 2016, accessed October 6, 2023, https://www.downtoearth.org.
in/news/water/anupam-mishra-the-man-who-dreamt-of-water-sufficient-
india-56578.

5. *Stepwells* are wells with a long corridor *of* steps descending to the water level,
with an ornamental as well as a cultural function.

6. Anupam Mishra Ted Talk accessed; https://www.ted.com/talks/anupam_
mishra_the_ancient_ingenuity_of_water_harvesting/transcript?subtitle=en;
10 mins 21 secs.

7. Based on M. Jain, 2021, Arvari case study, Natural and Nature Based
Features International Guidance, https://www.gov.uk/government/news/
natural-and-nature-based-features-international-guidance.

8. M. S. Rathore, 2003, "Community based management of groundwater
resources: A Case Study of Arwari River Basin," Institute of Development
Studies, Jaipur (Project Report submitted to British Geological Survey).

9. J. C. Glendenning, December 2010, "Evaluating the Impacts of Rainwater
Harvesting (RWH) in a Case Study Catchment: The Arvari River,
Rajasthan," Agricultural Water Management 98(2): 331–342.

10. Rathore.

11. R. N. Athavale, emeritus scientist at the National Geophysical Research
Institute/ Hyderabad.

12. Anupam Mishra on the image from Sita Bawdi in Bhilwara Rajasthan

Chapter 3

1. Wáng Lái-tōng, "Construction of dams according to the Heavenly Times and the Earthly Munificence," from multiple Chinese sources, translated by Ziwei Fan.
2. Wáng Lái-tōng, "On the Water Nature," from multiple Chinese sources, translated by Ziwei Fan.
3. "Taming the Floodwaters," China Heritage Project, The Australian National University, archived from the original on July 19, 2011, accessed March 14, 2023, http://www.chinaheritagenewsletter.org/features .php?searchterm=001_water.inc&issue=001.
4. D. Lee, "People's Party Animals," *Los Angeles Times*, February 8, 2006.
5. Dao de Jing, 1948, translated by Lin Yutang, Chapter 8 accessed online on May 27, 2023, https://terebess english/tao/yutang.html.hu/.

Chapter 4

1. "Reforesting Baran," The Flow Project, www.onepondfund.org, India, https://waterways.world/india/Re-ForestingBaran.pdf.
2. "Yacouba Sawadogo–Inspiration and Action," United Nations Environment Programme, accessed March 14, 2023, https://www.unep.org /championsofearth/laureates/2020/yacouba-sawadogo.
3. M. O'Connell, 2022, *Designing Regenerative Food Systems*, Hawthorn Press, p. 185.
4. Ernst Toch, 1977, *The Shaping Forces in Music: An Inquiry into the Nature of Harmony, Melody, Counterpoint, Form*, Dover Books, page 24.

Chapter 5

1. Y. Wada, P. H. van Beek, C. M. van Kempen, J. Reckman, S. Vasak, and M. Bierkens, October 26, 2010, "Global depletion of groundwater resources," *Hydrology and Land Surface Studies*, accessed May 23, 2023, https://agupubs .onlinelibrary.wiley.com/doi/full/10.1029/2010GL044571.
2. "Totnes flood defence scheme gets £3.8 million upgrade," 2018, Environment Agency, accessed May 28, 2023, https://www.gov.uk /government/news/totnes-flood-defence-scheme-gets-38- million-upgrade.
3. "Current Reservoir Storage," South West Water, accessed April 20, 2023. https://www.southwestwater.co.uk/environment/water-resources/current -reservoir-storages/.
4. "Estimating the Economic Costs of the 2015 to 2016 Winter Floods," 2016, Environment Agency, https://assets.publishing.service.gov.uk/government

/uploads/system/uploads/attachment_data/file/672087/Estimating_the
_economic_costs_of_the_winter_floods_2015_to_2016.pdf.

5. "Storm Desmond Case Study," A Level Geography, https://www
.alevelgeography.com/storm-desmond-case-study/.

6. P. Quinn, G. O'Donnell, A. Nicholson, M. Wilkinson, G. Owen, J. Jonczyk,
N. Barber, M. Hardwick, and G. Davies, 2013, "Potential Use of Runoff
Attenuation Features in Small Rural Catchments for Flood Mitigation,"
NFM RAF Report, https://research.ncl.ac.uk/proactive/belford
/newcastlenfmrafreport/reportpdf/June%20NFM%20RAF%20Report.pdf.

7. B. Cronin, February 29, 2016, "Keeping Pickering Flood Free," *New Civil
Engineer*, accessed May 29, 2023, https://www.newcivilengineer.com
/archive/keeping-pickering-flood-free-29-02-2016/.

8. G. Lean, January 3, 2016, "UK Flooding: How a Yorkshire town worked with
nature to stay dry," *Independent*, accessed May 29, 2023, https://www
.independent.co.uk/climate change/news/uk-flooding-how-a-yorkshire-flood
-blackspot-worked-with-nature-to-stay-dry a6794286.html.

9. Wilkinson, M. Quinn, P. Benson, I. Welton, P. (2010). Runoff Management:
Mitigation measures for disconnecting flow pathways in the Belford Burn
catchment to reduce flood risk. In: British Hydrological Society (Ed.), British
Hydrological Society International Symposium. 2010, Newcastle upon Tyne.

Chapter 6

1. Water School Africa, https://www.waterschools.org.

2. T. Pakenham, 1992, *The Scramble for Africa: White Man's Conquest of the Dark
Continent from 1876 to 1912*, Abacus Books, pp. 487–503.

3. Anna Brazier, 2020, "Harnessing Zimbabwe's indigenous knowledge for
a changing climate," Konrad Adenauer Stiftung, Harare, Zimbabwe, pp.
12–13.

4. SCOPE (*Schools and Colleges Permaculture*) and ReSCOPE (*Regional Schools and
Colleges Permaculture*) website, https://www.seedingschools.org/.

5. RIDEP website, https://www.ridepkenya.org.

6. M. E. Ellis, 2023, "Asbestos in Soil and Water," Mesotheolioma.net, accessed
December 23, 2023, https://mesothelioma.net/asbestos-in-soil-and-water/.

7. T. Pakenham, 1992, *The Scramble for Africa 1876-1912*, Abacus Books.

8. J. M. Coetzee, 1983, *The Life and Times of Michael K*, Ravan Press, page 184.

9. "Dr. Kenneth Wilson," Forever Sabah, https://www.foreversabah.org/team
-info/dr-kenneth-wilson.

10. I. Scoones, 2010, *Zimbabwe's Land Reform: Myths and Realities*, Weaver Press.

11. Ken Wilson, 2015, "Trees and Woodland Management: 2015 Statement of Program Rationale and Outcomes," accessed November 26, 2023, at https://muonde.org/category/trees-woodland-management/.

12. Scoones and Wilson, "Progress Report on Research and Development Activities in Zvishavane District, 1986–1987, page 3, https://zimbabweland .files.wordpress.com/2015/09/phiri-archive-4.pdf.

13. Ken Wilson, 2015, "Water Harvesting, Catchment Management and Farming," Statement of Program Rationale and Outcomes, accessed November 26, 2023, https://muonde.org/2015/03/01/water-harvesting-catchment-management-and-farming-2015-statement-of-program-rationale -and-outcomes/.

14. Scoones and Wilson, page 1.

15. Zephaniah Phiri, quoted in Scoones and Wilson, pages 3–4.

16. "Mazvihwa Indigeous Innovation Support Programme," Muonde Trust, page 1, https://www.muonde.org/wp-content/uploads/2013/03/Indigenous -Innovation-Support-Program.pdf.

17. Muonde Trust, https://muonde.org/the-work/.

18. Wilson, 2015

19. Wilson, 2015

20. Scoones, "Pfumvudza," https://zimbabweland.wordpress.com/tag/pfumvudza.

Chapter 7

1. NULC website, http://base.d-p-h.info/en/fiches/dph/fiche-dph-7066.html.

2. PORET website, https://www.poret.org/.

3. ZIMSOFF website, http://zimsofforum.org/.

4. Rob Sacco presentation, Water School Africa, waterways.world/africa.

5. Rob Sacco, personal interview.

6. Ibid.

7. PORET presentation, Water School Africa, waterways.world/africa.

8. Nelson Mudzwinga, 2021, presentation to Water School Africa, accessed December 23, 2023, https://www.youtube.com/watch?v=nCXQ1kMLYJU &t=581s.

9. T. Mahohoma, 2020, "Experiencing the Sacred," *Studia Historiae Ecclesiasticae* 46: 1, accessed November 28, 2023, http://www.scielo.org.za/scielo .php?script=sci_arttext&pid=S1017-04992020000100008.

10. Ibid.

11. Ibid.

Chapter 8

1. Emergency Response Coordination Centre (2019) "Tropical Cyclone Idai Impact Overview" accessed December 23, 2023, https://reliefweb.int/sites /reliefweb.int/files/resources/ECDM_20190318_TC_IDAI_update.pdf; "Darkness in the Wake of Idai," NASA (2019) accessed December 23, 2023 https://earthobservatory.nasa.gov/images/144743/darkness-in-the-wake-of-idai.

2. "Zimbabwe: Floods Flash Update No. 2," March 18, 2019, UN OCHA Office for the Coordination of Humanitarian Affairs, accessed December 2, 2023, https://reliefweb.int/report/zimbabwe/zimbabwe-floods-flash-update -no-2-18-march-2019.

3. Nick Davies, "Britain's Water Crisis," https://www.theguardian.com /environment/2015/oct/08/are-we-killing-our-rivers.

4. CELUCT website, www.celuct.org.

Chapter 9

1. M. Claussen, C. Kubatzki, V. Brovkin, A. Ganopolski, P. Hoelzmann, and H. Pachur, 1999, "Simulation of an abrupt change in Saharan vegetation in the Mid-Holocene," *Geophysical Research Letters* 26:14, accessed December 30, 2023, https://agupubs.onlinelibrary.wiley.com/doi/abs/10.1029/1999GL900494.

2. D. Wright, January 26, 2017, "Humans as Agents in the Termination of the African Humid Period," *Frontiers in Earth Science*, https://www.frontiersin .org/articles/10.3389/feart.2017.00004/full.

3. P. deMenocal and J. Tierney, 2012, "Green Sahara: African Humid Periods Paced by Earth's Orbital Changes," *Nature Education Knowledge* 3:10, p. 12.

4. "Over 14 billion USD raised for Great Green Wall to regreen the Sahel," UN Convention to Combat Desertification, https://www.unccd.int/news -stories/press-releases/over-14-billion-usd-raised-great-green-wall-regreen-sahel.

5. D. Todd, February 2023, "Independent Review of the Great Green Wall Accelerator," UN Convention to Combat Desertification, p. 12, accessed December 23, 2023, https://www.unccd.int/sites/default/files/inline-files /GGWA%20review%20final%20report%20formatted.pdf.

6. Todd, annex 1.

7. Todd, p. 27.

8. Tsuamba Borgou, personal interview.

9. Farmer Managed Natural Regeneration (FMNR) website, https://fmnrhub .com.au.

10. P. Cunningham and T. Abasse, 2005, "Reforesting the Sahel: Farmer Managed Natural Regeneration," published in A. Kalinganire, A. Niang, and

A. Kone, *Domestication des especes agroforestieres au Sahel: situation actuelle et perspectives*, ICRAF Working Paper, p. 2.

11. Cunningham and Abasse, p. 4.
12. Serving in Mission website, https://www.sim.org/home.
13. Cunningham and Abasse, p. 3.
14. FMNR website. https://fmnrhub.com.au/
15. H. Girard, 2009, "WE'GOUBRI, the sahelian bocage: an integrate approach for environment preservation and social development in sahelian agriculture (Burkina Faso)," *Field Actions Science Reports*, pp. 2, 33–39.
16. Terre Verte website, https://eauterreverdure.org/.
17. *The Man Who Stopped the Desert* website, http://www.1080films.co.uk/yacoubamovie/index.htm.
18. R. Belmin, December 27, 2023, "The zaï technique: how farmers in the Sahel grow crops with little to no water," The Conversation, accessed December 29, 2023, at https://theconversation.com/the-za-technique-how-farmers-in-the-sahel-grow-crops-with-little-to-no-water-220103.
19. D. Clavel, A. Barro, T. Belay, R. Lahmar, and F. Maraux, December 3, 2008, "Changements techniques et dynamique d'innovation agricole en Afrique Sahelienne: le cas du Zaï mécanisé au Burkina Faso et de l'introduction d'une cactée en Ethiopie," *Vertigo* 8:3.
20. P. deMenocal and J. Tierney, 2012, "Green Sahara: African Humid Periods Paced by Earth's Orbital Changes," *Nature Education Knowledge* 3:10, p. 12.

Chapter 10

1. People and Water website, https://peopleandwater.international/.
2. Peter Bunjak, "How to Create a Raingarden," The Flow Partnership, https://youtu.be/KcoO4UnfjYM?si=_Wf_n2J_n50fLzWR.
3. People and Water website, https://peopleandwater.international/.
4. M. Kravcik, J. Pokorny, J. Kohutiar, M. Kovac, and E. Toth, 2007, *Water for the Recovery of the Climate–A New Water Paradigm*, accessed May 30, 2023, http://www.waterparadigm.org/download/Water_for_the_Recovery_of_the_Climate_A_New_Water_Paradigm.pdf.
5. Estimates vary: see Y. Wada, P. H. van Beek, C. M. van Kempen, J. Reckman, S. Vasak, and M. Bierkens, October 26, 2010, "Global depletion of groundwater resources,"Hydrology and Land Surface Studies, accessed May 23, 2023, https://agupubs.onlinelibrary.wiley.com/doi/full/10.1029/2010GL044571.

6. Michal Kravcik, from The Flow Partnership, "How to Build a Raingarden" and "How to Make Wooden Dams," https://youtu.be/ULHDRyvpL0k?si= shqCi_P0klC-5Dra.

7. "Rain gardens," Royal Horticultural Society, https://www.rhs.org.uk /garden-features/rain-gardens.

Chapter 11

1. E. Heggy, J. Normand, E. Palmer, and A. Z. Abotalib, January 2022, "Exploring the nature of buried linear features in the Qatar peninsula: Archaeological and paleoclimatic implications," *ISPRS Journal of Photogrammetry and Remote Sensing* 183, pp. 210–227.

2. B. Ecclestone, J. Pike, and I. Harhash, 1981, *The Water Resources of Qatar and Their Development*, FAO.

3. Phillip Macumber, 2015, "Water Heritage in Qatar," Unesco World Heritage Convention, page 232, accessed June 8, 2023 (need access membership to Researchgate), https://www.researchgate.net/publication/304062991_Water_ Heritage_in_Qatar.

4. Phillip Macumber, "Water Heritage in Qatar" and "A geomorphological and hydrological underpinning for archaeological research in Northern Qatar," Vol. 41, Papers from the forty-fourth meeting of the Seminar for Arabian Studies held at the British Museum, London, 22 to 24 July 2010 (2011), pp. 187–200.

5. Y. E. Mohieldeen, E. A. Elobaid, and R. Abdalla, 2021, "GIS-based framework for artificial aquifer recharge to secure sustainable strategic water reserves in Qatar arid environment peninsula," *Science Reports* 11:18184, https://doi.org/10.1038/s41598-021-97593-w.

6. The Eemian interglacial was a period of increased global temperatures from 127,000 to 106,000 years ago, https://www.sciencedirect.com/topics/earth -and-planetary-sciences/eemian.

7. Phillip Macumber, "A geomorphological and hydrological underpinning for archaeological research in Northern Qatar," accessed June 8, 2023, https:// www.researchgate.net/publication/286761179_A_geomorphological_and_ hydro logical_underpinning_for_archae ological_research_in_Northern_ Qatar/download.

8. S. Aloui, A. Zghibi, A. Mazzoni, A. Elomri, and C. Triki, December 2023, "Groundwater resources in Qatar: A comprehensive review and informative recommendations for research, governance, and management in support of sustainability," *Journal of Hydrology: Regional Studies*, vol. 50.

9. Mott Macdonald, "Flood-proofing Doha's future," https://www.mottmac
    .com/article/76058/flood-proofing-dohas-future.

10. Ibid.

11. Y. E. Mohieldeen et al.

12. Aloui et al.

13. Ellen Kershner, "What Was the Dust Bowl?" https://www.worldatlas.com
    /articles/what-was-the-dust-bowl.html.

14. J. Prihantono, N. Adi, T. Nakamura, and K. Nadaoka, 2021, "The Impact
    of Groundwater Variability on Mangrove Greenness in Karimunjawa
    National Park based on Remote Sensing Study," 3rd International
    Conference on Maritime Sciences and Advanced Technology; IOP Conf.
    Series: Earth and Environmental Science 925, https://iopscience.iop.org/
    article/10.1088/1755-1315/925/1/012064#:~:text=Thus%2C%20these%20
    findings%20show%20that,and%20is%20vulnerable%20to%20drought.

15. M. Hayes, A. Jesse, N. Welti, B. Tabet, B. Lockington, and C. Lovelock,
    November 14, 2018, "Groundwater enhances above-ground growth in
    mangroves," *Journal of Ecology*, accessed January 23, 2024, https://
    besjournals.onlinelibrary.wiley.com/doi/full/10.1111/1365-2745.13105.

16. M. A. Shehadi, 2015, "Vulnerability of mangroves to sea level rise in Qatar:
    Assessment and identification of vulnerable mangroves areas, masters thesis,
    Qatar University.

17. H. Khader, March 23, 2023, "Mangroves in Qatar: Perspectives," *Environment
    Middle East Nature*, accessed October 2, 2023, https://www.ecomena.org
    /mangroves-in-qatar/.

18. Qatar Moments, 2022, "Plantation Programs in Qatar to Help Achieve Net-
    Zero Target by 2050," accessed October 2, 2023, https://www.qatarmoments
    .com/plantation-programs-in-qatar-to-help-achieve-netzero-target-by-2050
    -502354.html.

Chapter 12

1. "Thirsty Koala," DW News, https://www.youtube.com/
    watch?v=bwf9yQhYVrA.

2. Jérôme Fritel and P. Des Mazery, *Lords of Water*, film, 2019, accessed January
    16, 2024, https://www.youtube.com/watch?v=e5i35dNPGhs.

3. Ibid.

4. Ibid.

5. Ibid.

6. Ibid.

7. Ibid.

8. Greta Thunberg, 2019, "If World Leaders Choose to Fail Us, My Generation Will Never Forgive Them," *The Guardian*, accessed January 16, 2023, https://www.theguardian.com/commentisfree/2019/sep/23/world-leaders-generation-climate-breakdown-greta-thunberg.

9. Artlandish, "Aboriginal Dreamtime," at https://www.aboriginal-art-australia.com/aboriginal-art-library/aboriginal-dreamtime/.

10. Nutley, B (2022) Water and the Yarning Cycle; Australian Water Association accessed May 19, 2024, https://www.awa.asn.au/resources/latest-news/news/https/www.awa.asn.au/resources/latest-news/news/indigenous-water-values.

11. Bruce Pascoe, 2014, *Dark Emu*, Magabala Books.

12. R. Touma, 2023, "The Dark Emu Story: The Legacy—and Controversy—of Bruce Pascoe's Groundbreaking Book," *The Guardian*, accessed January 16, 2024, https://www.theguardian.com/film/2023/jul/18/dark-emu-story-bruce-pascoe-controversy-legacy-abc.

13. Pascoe, pp. 26–27.

14. T. L. Mitchell, 1839, *Three Expeditions into the Interior of Eastern Australia*, vol. 1, T. and W. Boone, publishers, p. 90.

15. R. Gerritsen, 2008, *Australia and the Origins of Agriculture*, Archaeopress, p. 33.

16. Charles Darwin, January 19, 1836, *The Beagle Diary*, accessed January 23, 2024, https://darwinbeagle.blogspot.com/2011/01/19th-january-1836.html.

17. J. Rickard, 2017, *Australia: A Cultural History*, Monash University Publishing, p. 48, accessed January 23, 2024, https://library.oapen.org/bitstream/id/774abcfa-c2d1-401f-a17a-987464a85366/645355.pdf.

18. Pascoe, p. 17.

19. E. Rolls, 1981, *A Million Wild Acres*, Nelson, p. 84.

20. Pascoe, p. 17.

21. *Lords of Water.*

22. *Lords of Water.*

23. *Lords of Water.*

24. "National Water Grid," Australian Government, https://www.nationalwatergrid.gov.au/about/water-in-australia.

25. Poelina, A., Taylor, K.S., Perdrisat, I., 2019. Martuwarra Fitzroy River Council: an Indigenous cultural approach to collaborative water governance. Australasian Journal of Environmental Management 0, 1–19; Martuwarra Fitzroy River Council: an Indigenous cultural approach to collaborative

water governance https://www.tandfonline.com/doi/abs/10.1080/1448 6563.2019.1651226; Poelina, A., 2020. A Coalition of Hope! A Regional Governance Approach to Indigenous Australian Cultural Wellbeing, in: Campbell, A., Duffy, M., Edmondson, B. (Eds.), Located Research: Regional Places, Transitions and Challenges. Springer, Singapore, pp. 153–180. https://link.springer.com/chapter/10".1007/978-981-32-9694-7_10.

26. Indigenous Knowledge Institute, 2024, "Indigenous Voices in Water," University of Melbourne, accessed January 16, 2024, https:// indigenousknowledge.unimelb.edu.au/curriculum/resources/indigenous -voices-in-water.

27. The Convention on Wetlands, Ramsar website https://www.ramsar.org.

28. Australian government, 2024, "Wetlands and Indigenous Values," accessed January 16, 2024, https://www.dcceew.gov.au/water/wetlands/publications /factsheet-wetlands-indigenous-values.

29. A. L. Payne, I. W. Watson, and P. E. Novelly, 2004, "Spectacular recovery in the Ord River catchment," Department of Primary Industries and Regional Development, Western Australia, Perth, Report 17/2004.

30. Tarwyn Park Training website, https://www.tarwynparktraining.com.au/ about-us.

31. B. Cheshire, 2015, "Iconic rural property Tarwyn Park should be given heritage listing, says study by Newcastle University," ABC News, accessed January 16, 2024, https://www.abc.net.au/news/2015-05-04/tarwyn-park-should-be-given-heritage-listing,-says-newcastle-uni.

32. Regenerative Landscapes Australia, "Soaking up Australia's Drought," https://www.rlaustralia.com.au/soaking-up-australias-drought/.

33. Lords of Water.

34. Lords of Water.

35. UN Resolution 64/292, www.un.org.

Chapter 13

1. Village Udguwan, block Talbehat, district Lalitpur, Bundelkhand, India.

2. Parmarth Social Service Organization (PSSS), https://parmarthindia.com/.

3. Village Hanauta, block Talbehat, district Lalitpur, Bundelkhand, India.

4. Village Rund Ballora, Durgapur, block Babina, district Jhansi, Bundelkhand, India.

5. A gram panchayat ("village council") is a basic governing institution in Indian villages, (roughly three villages form one gram panchayat).

Chapter 14

1. Ben Goldfarb, 2018, *Eager: The Surprising, Secret Life of Beavers and Why They Matter*, Chelsea Green, p. 59.

2. "Meet the Beavers of Knapdale," Forestry and Land Scotland, https://forestryandland.gov.scot/blog/meet-the-beavers-of-knapdale.

3. Scottish Beaver Trial, https://www.nature.scot/professional-advice/protected-areas-and-species/protected species/protected-species-z-guide/beaver/scottish-beaver-trial.

4. "Wild beavers seen in England for first time in centuries," The Guardian, February 27, 2014, accessed October 2, 2023, https://www.theguardian.com/environment/2014/feb/27/wild-beavers-england-devon-river.

5. "'First' sighting of wild beavers in England for centuries," BBC News, February 27, 2024, accessed October 2, 2023, https://www.bbc.co.uk/news/uk-england-devon-26365127.

6. "Correspondence Which Discusses Beavers in British Waterways," DEFRA, accessed October 2, 2023, https://assets.publishing.service.gov.uk/government/uploads/system/uploads/attachment_data/file/361557/6739_Response_-_redacted_version_for_publication__2__amended.pdf.

7. Natural England, https://www.gov.uk/government/organisations/natural-england.

8. "Science and Evidence Report," River Otter Beaver Trial, https://www.exeter.ac.uk/media/universityofexeter/research/microsites/creww/riverottertrial/ROBT__Science_and_Evidence_Report_2020_(ALL).pdf.

9. D. Butler and G. Malanson, October 2005, "The geomorphic influences of beaver dams and failures of beaver dams," *Geomorphology* 71:1–2, Elsevier, pp. 1, 48–60.

10. Butler and Malanson, p. 1.

11. K. Samurović, 2021, "Building It Back—Beaver Reintroductions across the World," *Geography Realm*, accessed January 23, 2024, https://www.geographyrealm.com/building-it-back-beaver-reintroductions-across-the-world/.

12. Wild Trout Trust, "Beavers Beneath the Surface," https://www.wildtrout.org/content/beavers-benefits-beneath-the-surface.

13. "Landmark decision gives wild beavers permanent right to remain in England," BBC Wildlife Magazine, https://www.discoverwildlife.com/news/landmark-decision-gives-wild-beavers-permanent-right-to-remain-in-england.

Chapter 15

1. Felipe Medina, 2023, "Origen Circular," Pasos del Agua (Steps of Water) website, accessed December 23, 2024, https://origencircular.com/agua/.
2. SIWI, Rajendra Singh, "The Water Man of India receives Stockholm Water Prize" Singh, R. (2015) The Water Man of India receives Stockholm Water Prize, SIWI (2015) accessed December 2023, https://siwi.org/latest/the-water-man-of-india-receives-stockholm-water-prize/.
3. Scoones, "The Water Harvester, Zephaniah Phiri" Scoones, I (2015) "The Water Harvester, Zephaniah Phiri has Died" accessed online on 23rd December 2023 at https://zimbabweland.wordpress.com/2015/09/02/the-water-harvester-zephaniah-phiri-has-died.
4. "Claudia Steiner," Reserva Guatoc, https://www.reservaguatoc.com/nosotros.
5. "Senderos del Agua," Reserva Guatoc, https://www.reservaguatoc.com/portfolio-1/project-one-f5w4d-xmlx4.
6. There is an old story that the Mother of Water was a big snake with a crest and a gizzard that guarded the creeks and the wellsprings—a mediator between the two worlds. The story is that when this snake is killed, the wellspring dies. A lot of contemporary stories relate to community members "seeing" the snake. This snake is still part of the local folklore, even though Barichara is now a Catholic town, showing the importance of this symbolism.
7. Joe Brewer, "Walking the Path of Regeneration," https://medium.com/@joe_brewer/walking-the-path-of-regeneration-af56e6e54c0c.
8. "A Waste Water Treatment Plant (WWTP) for Barichara," 2022, Anti-Corruption Institute, accessed December 23, 2023, https://www.estudiosanticorrupcion.org/en/a-waste-water-treatment-plant-wwtp-for-barichara-we-support-the-local-communitys-popular-action-in-the-municipality/.
9. "Eight years after the contract was signed, Barichara is still without an aqueduct," November 13, 2019, Vanguard, accessed December 23, 2023, https://www-vanguardia-com.translate.goog/politica/ocho-anos-despues-de-firmado-el-contrato-barichara-sigue-sin-acueducto-AI1659323?_x_tr_sl=es&_x_tr_tl=en&_x_tr_hl=en&_x_tr_pto=sc.
10. Thomas Berry, 2009, *The Christian Future and the Fate of Earth*, edited by Mary Evelyn Tucker and John Grim, Orbis Books.
11. The sentiment for this popular saying is found in many cultures. "They have forsaken me, the spring of living water, and have dug their own cisterns, broken cisterns that cannot hold water." Jeremiah 2:13.

12. Felipe Medina, "Pasos del Agua, Steps of Water," https://origencircular.com/agua/.

13. Cathy Holt, 2022, "Barichara December Solstice," Earth and Us, accessed December 23, 2023, https://cathyholt.medium.com/earth-us-d4e52329ea35.

Chapter 16

1. B. Eichengreen, 2019, *Globalizing Capital: A History of the International Monetary System*, 3rd ed., Princeton University Press, p. 130, https://delong.typepad.com/files/eichengreen-globalizing.pdf.

2. Eichengreen, p. 144.

3. Vandana Shiva, 1991, "The Green Revolution in the Punjab," *The Ecologist* 21 (2): 57–60. Accessed May 27, 2023, https://www.yumpu.com/en/document/read/38148491/the-green-revolution-in-the-punjabpdf.

4. "What is the national debt?" 2023, Fiscal Data US Treasury, accessed May 23, 2023, https://fiscaldata.treasury.gov/americas-finance-guide/national-debt/#the-growing-national-debt.

5. Y. Wada, L. van Beek, C. van Kempen, W. Reckman, S. Vasak, and M. Bierkens, "Global depletion of groundwater resources," October 26, 2010, *Hydrology and Land Surface Studies*, accessed May 23, 2023, https://agupubs.onlinelibrary.wiley.com/doi/full/10.1029/2010GL044571.

6. S. Rost, D. Gerten, A. Bondeau, W. Lucht, J. Rohwer, and S. Schaphoff, 2008, "Agricultural green and blue water consumption and its influence on the global water system," *Water Resources Research* 44: 9, https://agupubs.onlinelibrary.wiley.com/doi/10.1029/2007WR006331.

7. C. J. Vörösmarty, C. Lévêque, and C. Revenga, 2005, "Fresh Water," in *Ecosystems and Human Well-being: Current State and Trends*, ed. R. Hassan et al., Island Press, pp. 167– 207.

8. US Geological Survey, 2018, "Groundwater Decline and Depletion," accessed May 23, 2023, https://www.usgs.gov/special-topics/water-science-school/science/groundwater-decline-and-depletion?qt science_center_objects=0#qt-science_center_objects.

9. Global Wetland Outlook, Ramsar Wetlands Convention, 2021, accessed May 23, 2023, https://www.global-wetland-outlook.ramsar.org/report-1.

10. N. Shukla, "Water Is Now Being Traded as a Commodity Amid Fears of Scarcity," accessed May 30, 2023, https://earth.org/water-trade/; P. Almendros, 2020, "The Future of Water Is Traded in the Stock Exchange," *Smart Water Magazine*, December 9, 2020, accessed May 30, 2023, https://

smartwatermagazine.com/news/smart-water-magazine/future-water-traded
-stock-exchange; J. McWhinney, December 31, 2021, "Water Investments:
How to Invest in Water," *Investopedia*, accessed May 30, 2023, https://www
.investopedia.com/articles/06/water.asp.

11. P. Greenfield, January 18, 2023, "More than 90% of rainforest carbon offsets
by biggest certifier are worthless," The Guardian, accessed May 30, 2023,
https://www.theguardian.com/environment/2023/jan/18/revealed
-forest-carbon-offsets-biggest-provider-worthless-verra-aoe.

12. M. Rodell, "Satellites Unlock Secret to Northern India's Vanishing Water,"
August 12, 2009, accessed July 8, 2023, https://www.jpl.nasa.gov
/news/satellites-unlock-secret-to-northern-indias-vanishing-water.

13. "The Water Man of India receives Stockholm Water Prize," 2015,
Stockholm International Water Institute (SIWI), accessed December 23,
2023, https://siwi.org/latest/the-water-man-of-india-receives-stockholm
-water-prize/.

14. Water Schools Africa, www.waterschools.org.

15. Lorenzo Marsili, January 9, 2024, "War gave us the Red Cross. Now climate
disaster means we need a Green Cross too," *The Guardian*, accessed January
10, 2024, https://www.theguardian.com/commentisfree/2024/jan/09/war-
red-cross-green-climate-disasters.

16. Yvon Chouinard, 2023, Patagonia, accessed May 3, 2023, https://
eu.patagonia.com/gb/en/ownership/.

Conclusion

1. Nick Miroff, March 7, 2015, "Havana, from on High," *Washington Post*,
https://www.washingtonpost.com/sf/world/2015/03/07/on-havanas-
rooftops-a-secret-world/.

2. Mickey Ellinger, 2010, "Urban Agriculture in Cuba," Reimagine 17: 2,
https://www.reimaginerpe.org/17-2/ellinger.

3. *Return of the River*, film website, http://www.elwhafilm.com/.

4. Water Schools, www.waterschools.org.

5. Yvon Chouinard, 2023, Patagonia, accessed May 3, 2023, https://
eu.patagonia.com/gb/en/ownership/.

BIBLIOGRAPHY

Almendros, P. (2020) "The Future of Water Is Traded in the Stock Exchange." *Smart Water Magazine*, 9 December 2020. Accessed 30 May 2023. https://smartwatermagazine.com/news/smart-water-magazine/future-water-traded-stock-exchange.

Aloui, S., Zghibi, A., Mazzoni, A., Elomri A., and Triki, C. (2023) "Groundwater resources in Qatar: A comprehensive review and informative recommendations for research, governance, and management in support of sustainability." *Journal of Hydrology: Regional Studies*. Volume 50, December 2023, 101564.

Barnes, M., Zhang, Q., Robeson, S., Young, L., Burakowski, E., Oishi, A., Stoy, P., Katul, G., Novick, K. (2024) "Century of Reforestation Reduced Anthropogenic Warming in the Eastern United States." *Earth's Future*. 13 February 2024, Vol 12 Issue 2.

Berry, T. (2009) *The Christian Future and the Fate of Earth*. Edited by Mary Evelyn Tucker and John Grim. Maryknoll, Orbis Books.

Brazier, A. (2020) *Harnessing Zimbabwe's Indigenous Knowledge for a Changing Climate*. Konrad Adenauer Stiftung, Harare, Zimbabwe.

Butler, D. and Malanson, G. [2005] "The geomorphic influences of beaver dams and failures of beaver dams." *Geomorphology*, Volume 71, Issues 1–2, October, pp. 48–60, Elsevier.

Chitongo, L., Tagarirofa, J., Chazovachii, B., and Marango, T. (2019) "Gendered Impacts of Climate Change in Africa: The Case of Cyclone Idai, Chimanimani, Zimbabwe, March 2019." *The Fountain–Journal of Interdisciplinary Studies*, Volume 3, Issue 1, Nov–Dec 2019.

Claussen, M., Kubatzki, C., Brovkin, V., Ganopolski, A., Hoelzmann, P., and Pachur, H. (1999) "Simulation of an Abrupt Change in Saharan Vegetation in the Mid-Holocene." *Geophysical Research Letters*, Volume

26; Issue 14. Accessed 30 December 2023 https://agupubs.onlinelibrary.
wiley.com/doi/abs/10.1029/1999GL900494.

Clavel, D., Barro, A., Belay, T., Lahmar, R., and Maraux, F. (2008)
"Changements techniques et dynamique d'innovation agricole en
Afrique Sahelienne: le cas du Zaï mécanisé au Burkina Faso et de
l'introduction d'une cactée en Ethiopie." *Vertigo*, Volume 8, Numéro 3,
December.

Coetzee, J. M. (1983) *The Life and Times of Michael K.* Ravan Press.

Cronin, B. (2016) "Keeping Pickering Flood Free." *New Civil Engineer*. 29
February. Access 29 May 2023. https://www.newcivilengineer.com/
archive/keeping-pickering-flood-free-29-02-2016/.

Cunningham, P., and Abasse, T. (2005) "Reforesting the Sahel: Farmer
Managed Natural Regeneration." Published in A. Kalinganire, A. Niang,
and A. Kone, (2005) *Domestication des especes agroforestieres au Sahel:
situation actuelle et perspectives.* ICRAF Working Paper.

Dao de Jing. Translated by Yutang Long (1948). Chapter 8 accessed online, 27
May 2023. https://terebess.hu/english/tao/yutang.html.

deMenocal, P., and Tierney, J. (2012) "Green Sahara: African Humid Periods
Paced by Earth's Orbital Changes." *Nature Education Knowledge* 3(10):12.

Denyer, K., Akoijam, Y., Ali, K. M., Khurelbaatar, S., Oviedo, G., and
Young, L. (2018) *Learning from Experience: How Indigenous Peoples and
Local Communities Contribute to Wetland Conservation in Asia and Oceania.*
Ramsar Convention Secretariat. Accessed 16 January 2024. https://www.
ramsar.org/sites/default/files/learning_from_experience_march_2018.
pdf.

Down to Earth. (2016) "Anupam Mishra: The Man Who Dreamt
of Water Self-Sufficiency in India." Accessed 6 October
2023. https://www.downtoearth.org.in/news/water/
anupam-mishra-the-man-who-dreamt-of-water-sufficient-india-56578.

Ecclestone, B., Pike, J., and Harhash, I. (1981) "The Water Resources of Qatar
and Their Development: Water Resources and Agricultural Development
Project" FAO funds-in-trust technical report no. 5: 2 vols. Food and
Agriculture Organization of the United Nations.

Engel, M., Rückmann, S., Drechsler, P., Brill, D., Opitz, S., Fassbinder, J. W.,
Pint, A., Peis, K., Wolf, D., Gerber, C., Pfeiffer, K., Eichmann, R., and
Brückner, H. (2020) "Sediment-Filled Karst Depressions and *Riyad*—Key
Archaeological Environments of South Qatar." *E&G Quaternary Sci. J.*,
68, 215–236.

Environment Agency (2018). "Estimating the Economic Costs of the 2015 to 2016 Winter Floods." https://assets.publishing.service.gov.uk/government/uploads/system/uploads/attachment_data/file/672087/Estimating_the_economic_costs_of_the_winter_floods_2015_to_2016.pdf.

Environment Agency (2018). "Totnes Flood Defence Scheme Gets £3.8 Million Upgrade." Accessed 28 May 2023. https://www.gov.uk/government/news/totnes-flood-defence-scheme-gets-38-million-upgrade.

Environment Agency (2022). "Natural Flood Management Programme: Evaluation Report." Accessed 15 January 2024. https://www.gov.uk/government/publications/natural-flood-management-programme-evaluation-report/natural-flood-management-programme-evaluation-report.

Frieden, M. (2019) "The First 6 Days After Cyclone Idai in Zimbabwe." Voices from the Field, *Medecins Sans Frontieres.* https://www.msf.org/cyclone-idai-zimbabwe-first-six-days.

Fritel, J. and Des Mazery, P. (2019) *Lords of Water: How Banks Are Profiting Off of the Water Crisis."* Earth Stories. Accessed 16 January 2024. https://www.youtube.com/watch?v=e5i35dNPGhs.

Gerritsen, R. (2008) *Australia and the Origins of Agriculture.* Archaeopress.

Girard, H. (2009) "WÉGOUBRI, the Sahelian Bocage: An Integrate Approach for Environment Preservation and Social Development in Sahelian Agriculture (Burkina Faso)." *Field Actions Sci. Rep.,* 2, 33–39.

Glendenning, J.C. (2010) Evaluating the impacts of Rainwater Harvesting (RWH) in a case study catchment: The Arvari River, Rajasthan, December 2010 Agricultural Water Management 98(2):331–342.

Hayes, M., Jesse, A., Welti, N., Tabet, B., Lockington, B., and Lovelock, C. (2018) "Groundwater Enhances Above-Ground Growth in Mangroves." *Journal of Ecology,* 14 November 2018. Accessed 23 January 2024. https://doi.org/10.1111/1365-2745.13105

Goldfarb, B. (2018) *Eager: The Surprising, Secret life of Beavers and Why They Matter.* Chelsea Green.

Heggy, E., Normand, J., Palmer, E., and Abotalib, A. Z. (2022) "Exploring the Nature of Buried Linear Features in the Qatar Peninsula: Archaeological and Paleoclimatic Implications." *ISPRS Journal of Photogrammetry and Remote Sensing,* Volume 183, January 2022, 210–227. https://doi.org/10.1016/j.isprsjprs.2021.10.007.

IB Environmental Science and Systems. (2023) "Transfer vs. Transformation." Accessed 30 May 2023. https://sites.google. com/site/vsenvironmentalsciencesystems/1-systems-models/ transfer-vs-transformation.

Jain, M. (2021) "Arvari Case Study" in *International Guidelines on Natural and Nature-Based Features for Flood Risk Management.* https://www.gov.uk/government/news/ natural-and-nature-based-features-international-guidance.

Kerschner, E. (2020) "What Was the Dust Bowl?" *World Atlas.* Accessed 23 January 2023. https://www.worldatlas.com/articles/what-was-the-dust-bowl.html.

Khader, H. (2023) "Mangroves in Qatar: Perspectives." EcoMENA: Echoing Sustainability in MENA. Accessed 2 October 2023. https://www. ecomena.org/mangroves-in-qatar/.

Kravcik, M., Pokorny, J., Kohutiar, J.. Kovac, M. Toth, E. (2007) *Water for the Recovery of the Climate– A New Water Paradigm.* Accessed 30 May 2023. http://www.waterparadigm.org/download/Water_for_the_Recovery_of_ the_Climate_A_New_Water_Paradigm.pdf.

Lean, G. (2016) "UK Flooding: How a Yorkshire Town Worked with Nature to Stay Dry." *Independent.* 3 January 2016. Accessed 29 May 2023. https://www.independent.co.uk/climate-change/news/uk-flooding-how-a-yorkshire-flood-blackspot-worked-with-nature-to-stay-dry-a6794286.html.

Lee, D. (2006) "People's Party Animals." 8 February 2006. *Los Angeles Times.*

Macumber, P. (2015) "Water Heritage in Qatar." UNESCO *World Heritage Convention.* 3 August 2015. Accessed 8 June 2023. https://www. researchgate.net/publication/304062991_Water_Heritage_in_Qatar.

Macumber, P. (2011) "A Geomorphological and Hydrological Underpinning for Archaeological Research in Northern Qatar." *Proceedings of the Seminar for Arabian Studies,* vol. 41, 187–200. Accessed 8 June 2023. https://www. researchgate.net/publication/286761179_A_geomorphological_and_ hydrological_underpinning_for_archaeological_research_in_Northern_ Qatar.

Mahohoma, T. (2020) "Experiencing the Sacred." *Studia Historiae Ecclesiasticae* vol.46 n.1. Accessed 28 November 2023. http://www.scielo.org.za/scielo. php?script=sci_arttext&pid=S1017-04992020000100008.

Mati, B. M. (2005) "Overview of Water and Soil Nutrient Management under Smallholder Rain-fed Agriculture in East Africa." International Water

Management Institute, Working Paper 105. https://www.iwmi.cgiar.org/
Publications/Working_Papers/working/WOR105.pdf

McWhinney, J. (2021) "Water Investments: How to Invest in Water."
Investopedia. 31 December 2021. Accessed 30 May 2023. https://www.
investopedia.com/articles/06/water.asp.

Mishra, A. (n.d.) *The Radiant Raindrops of Rajasthan (Rajasthan Ki Rajat
Boonden)*. Translated by Maya Jani. Accessed 5 October 2023. https://
www.arvindguptatoys.com/arvindgupta/anupam.pdf

Mishra, A. (1993) *The Ponds Are Still Relevant (Aaj Bhi Khare Hain Talaab)*.
Gandhi Peace Foundation. Accessed 5 October 2023. https://admin.
indiawaterportal.org/sites/default/files/iwp2/the_ponds_are_still_
relevant_2016.pdf.

Mitchell, T. (1839) *Three Expeditions into the Interior of Eastern Australia*. T and
W Boone

Mohieldeen,Y., Elobaid, E., and Abdalla, R. (2021) "GIS-Based Framework
for Artificial Aquifer Recharge to Secure Sustainable Strategic Water
Reserves in Qatar Arid Environment Peninsula." *Scientific Reports* 11:
18184. Accessed 8 June 8 2023. https://www.nature.com/articles/
s41598-021-97593-w.

NASA Earth Observatory (2019). "Darkness in the Wake of Idai." Accessed
23 December 2023. https://earthobservatory.nasa.gov/images/144743/
darkness-in-the-wake-of-idai.

Niyogi, D. G. (2021) "Women Revive Ponds for Water
Security in Bundelkhand." *Mongabay*. Accessed 6
October 2023. https://india.mongabay.com/2021/01/
women-revive-ponds-for-water-security-in-bundelkhand/.

O'Connell, M. (2022) *Designing Regenerative Food Systems*. Hawthorn Press.

Pakenham, T. (1991) *The Scramble for Africa: 1876–1912*. Abacus.

Pascoe, B. (2014) *Dark Emu: Black Seeds: Agriculture or Accident?* Magabala
Books.

Payne, A. L., Watson, I. W., and Novelly, P. E. (2004) *Spectacular Recovery in
the Ord River Catchment*. Department of Primary Industries and Regional
Development, Western Australia, Perth. Report 17/2004. https://library.
dpird.wa.gov.au/cgi/viewcontent.cgi?article=1005&context=misc_pbns.

Poelina, A., Taylor, K. S., and Perdrisat, I. (2019) "Martuwarra Fitzroy River
Council: An Indigenous Cultural Approach to Collaborative Water
Governance." *Australasian Journal of Environmental Management* 29, 1–19.
https://doi.org/10.1080/14486563.2019.1651226.

Poelina, A. (2020) "A Coalition of Hope! A Regional Governance Approach to Indigenous Australian Cultural Wellbeing" in *Located Research: Regional Places, Transitions, and Challenges*, eds. Campbell, A., Duffy, M., and Edmondson, B. Palgrave Macmillan, Singapore, 153–180. https://doi.org/10.1007/978-981-32-9694-7_10.

Prihantono, J., Adi, N. S., Nakamura, T., and Nadaoka, K. (2021) "The Impact of Groundwater Variability on Mangrove Greenness in Karimunjawa National Park Based on Remote Sensing Study." *IOP Conf. Ser.: Earth Environ. Sci.* 925 012064. https://iopscience.iop.org/article/10.1088/1755-1315/925/1/012064.

Quinn, P., O'Donnell, G., Nicholson, A., Wilkinson, M., Owen, G., Jonczyk, J., Barber, N., Hardwick, M. and Davies, G. (2013) "Potential Use of Runoff Attenuation Features in Small Rural Catchments for Flood Mitigation." NFM RAF Report.

Quinn, P. (2023) "Catchment System Engineering." Presentation to Water School Africa. Accessed 28 May 2023. https://www.youtube.com/watch?v=ykIm5y3-tac.

Rathore, M. S. (2003) "Community Based management of Ground Water Resources: A Case Study of Arwari River Basin." Institute of Development Studies, Jaipur (Project Report submitted to British Geological Survey).

Rolls, E. (1981) *A Million Wild Acres*. Nelson.

Rost, S., Gerten, D., Bondeau, A., Lucht, W., Rohwer, J., and Schaphoff, S. (2008) "Agricultural Green and Blue Water Consumption and Its Influence on the Global Water System." *Water Resources Research* 44, W09405, https://doi.org/10.1029/2007WR006331.

Scoones, I. (2010) *Zimbabwe's Land Reform: Myths and Realities*. Weaver Press.

Scoones, I., and Wilson, K. (1987) "Progress Report on Research and Development Activities in Zvishavane District, 1986–1987." Accessed 26 November 2023. https://zimbabweland.wordpress.com/wp-content/uploads/2015/09/phiri-archive-4.pdf.

Shiva, V. (1991) "The Green Revolution in the Punjab." *The Ecologist* 21 (2): 57–60. Accessed 27 May 2023. https://www.yumpu.com/en/document/read/38148491/the-green-revolution-in-the-punjabpdf.

SIWI. (2015) "The Water Man of India Receives Stockholm Water Prize." Stockholm Water Prize Laureates 2015. Accessed 25 January 2023. https://siwi.org/stockholm-water-prize/laureates/2015-rajendra-singh.

Toch, E. (1977) *The Shaping Forces in Music: An Inquiry into the Nature of Harmony, Melody, Counterpoint, and Form.* Dover Books.

Todd, D. (2023) "Independent Review of the Great Green Wall Accelerator." United Nations Convention to Combat Desertification. Accessed 23 December 2023. https://www.unccd.int/sites/default/files/inline-files/GGWA%20review%20final%20report%20formatted.pdf.

Vörösmarty, C. J., Lévêque, C., and Revenga, C. (2005) "Fresh Water" in *Ecosystems and Human Well-being: Current State and Trends,* ed. R. Hassan et al., 167– 207, Island, Washington, D. C. https://www.millenniumassessment.org/documents/document.276.aspx.pdf.

Wada, Y., van Beek, L. P. H., van Kempen, C. M., Reckman, J. W. T. M., Vasak, S., and Bierkens, M. F. P. (2010) "Global Depletion of Groundwater Resources." *Hydrology and Land Surface Studies* 26 October 2010. Accessed 23 May 2023. https://doi.org/10.1029/2010GL044571.

Wáng Lái-tōng. (1723–1799) *Construction of Dams Cccording to the Heavenly Times and the Earthly Munificence.* Sourced by Ziwei Fan.

Wáng Lái-tōng. (1723–1799) *On the Water Nature.* Sourced by Ziwei Fan.

Wilkinson, M., Quinn, P., Barber, N., Jonczyk, J. (2014) "A Framework for Managing Runoff and Pollution in the Rural Landscape Using a Catchment Systems Engineering Approach." *Science of the Total Environment* 468–469, 1245–1254. https://doi.org/10.1016/j.scitotenv.2013.07.055.

Wilson, K. (2015a) "Trees and Woodland Management: 2015 Statement of Program Rationale and Outcomes." https://muonde.org/2015/03/01/trees-and-woodland-management-2015-statement-of-program-rationale-and-outcomes/.

Wilson, K. (2015b) "Water Harvesting, Catchment Management and Farming: 2015 Statement of Program Rationale and Outcomes." Accessed 26 November 2023. https://muonde.org/2015/03/01/water-harvesting-catchment-management-and-farming-2015-statement-of-program-rationale-and-outcomes/

Wright, D. K. (2017) "Humans as Agents in the Termination of the African Humid Period." *Frontiers in Earth Science,* 26 January 2017, Sec. Quaternary Science, Geomorphology and Paleoenvironment, Volume 5. https://doi.org/10.3389/feart.2017.00004.

SUBJECT INDEX

5 Percent Future (UK), The, 78

Ancient Water Wisdom (China), 32

Approach (Colombia, South America), The, 220

Arriving into the Present (Africa), 98

Baoping Kou or Bottle-Neck Channel (宝瓶口) (China), The, 38

Belford Burn, United Kingdom (Small Catchment), 63

Blue Water Abstraction, 237

Blueprint for Renewal (Qatar, Middle East), 170

Breaking the Taboos (Bundelkhand, India), 195

Building a Rain Garden (Slovakia), 149

Can Anyone Spot Them? (USA/ UK), 210

Case Study: Arvari River Watershed (Alwar, Rajasthan, India), 28

Collective Action (Australia), 185

Colonialism (Colombia, South America), 225

Community Traditional Wisdom, 31

Conversing with the Government (Colombia, South America), 224

Dearth of Indigenous Care (Australia), The, 182

Decline of the Commons (Australia), The, 179

Desertification and Reforestation, 57

Design into Practice (UK), 70

Design Template for Large Woody Debris and Offline Ponds (UK), A, 81

Elements of the Design (China), 36

Exercises (Conclusion), 257–259

Falling Away from Tradition (Qatar, Middle East), 162

Fast Forward to the Modern Era, 236

Feishayan or Flying Sand Weir (飞沙堰) (China), The, 38

Financial Capital and Natural Capital, 234

Forgotten Language: Shifting Sands Down the Ages, A, 6

Forgotten Word, A, 45

Founding a Local Water Bank, 241

Friends of Water Together (Bundelkhand, India), 197

From Apathy to Awareness, 5

Fundamental Right (Australia), A, 184

Gardens Blooming in the Rain
(Slovakia), 156

Gentrification (Colombia, South
America), 228

Guardian of the Catchment: Silas
(Zimbabwe, Africa), 120

Guardian of the Community: John
Strong (Zimbabwe, Africa), 125

Guardian of the Trees: Godfrey
(Zimbabwe, Africa), 119

Guardians of the Spirit: Ancestors
and Forest Spirits (Zimbabwe,
Africa), 124

High Pasture Regeneration
(Zimbabwe, Africa), 127

How Does One Calculate the
Prosperity and Life That This
Water Bank Enables? 242

Hydrologists or Pests? (Beavers)
(USA/ UK), 208

"Imagine the Creek Running Clean"
(Colombia, South America),
229

Impacting the Water Cycle, 57

In Foolish Pursuit of a Hat (Beavers)
(USA/ UK), 213

In the Imagination (Qatar, Middle
East), 167

Indigenous Wisdom (Australia), 175

Intentional Recharge (Africa), 101

Investing in the Future, 246

Irrigation System (China), The, 34

Is Higher Runoff the Greatest Threat
to the Future? (UK), 61

Jal Sahelis (Bundelkhand, India),
The, 198

Key Components of the Water Cycle,
56

Landmark Decision Gives Wild
Beavers Permanent Right to
Remain in England (USA/
UK), 216

Language of Water as Taught by
Mangu Kaka (India), The, 22

Local Action's Impact on the Water
Cycle, 47

Man Who Stopped the Desert:
Farmer Yacouba Sawadogo
(Burkino Faso, Africa), The, 143

Mangrove Restoration: Signs of
Hope (Qatar, Middle East), 169

Modern Methods (Qatar, Middle
East), 165

Modern Times (Colombia, South
America), 226

Nature's Principles Uncovered by
Anupam Mishra (India), 16

Nine Simple Yet Profound Lessons
Rajendra Singh Learned on
His Journey with Water Revival
(India), 26

Nyahode Union Learning Centre
(Zimbabwe, Africa), 105

Oil and Water (Qatar, Middle East),
163

Participatory Organic Research
and Extension Training Trust
(Zimbabwe, Africa), 108

Past Informs the Present (Colombia,
South America), The, 225

Pathways to Change (Zimbabwe,
Africa), 129

Phiri's Successor: The Muonde
Trust, 93

Postscript (Colombia, South
America), 233

Potential Pathway, A, 247

Precolonialism (Colombia, South America), 225

Putting Principles into Practice (India), 19

Registering in a Beaver Class (USA/ UK), 209

Remembering the Language of Water, 53

Resolving the Impasse (Colombia, South America), 229

Restoring the Water Balance, 8

Returning to an Old Banking System, 246

Revitalized Identity (Africa), A, 103

River Dart, United Kingdom (Medium-Size Catchment), 65

River Eden, United Kingdom (Large Catchment), 67

Saseri Water Bank Branch (India), The, 239

Shareholders Meeting, 243

Shashe Farms (Zimbabwe, Africa), 110

Stimulating Action (Colombia, South America), 219

Strengthening the Muscle: Natural Sequence Farming and Peter Andrews (Australia), 184

Tale of Two Villages (India), A, 50

Tentative Step (Colombia, South America), A, 223

Three Paths to Desertification (Burkino Faso, Africa), 131

Three Routes to Transformation (Burkino Faso, Africa), 133

Three Steps to Revival (Burkino Faso, Africa), 144

Traditional Methods of Managing Rainwater (Qatar, Middle East), 159

Transformation Remembered, 58

Translating Thought into Action, 12

Tribute to the Beavers (USA/ UK), 214

Turning the Key to Transformation (USA/ UK), 212

Two Ways of Farming, 92

Vessel Appears, The (Sanjay Singh and PSSS) (Bundelkhand, India), 195

Water Banks of the World, The, 238

Water Symphony, A, 59

Weighing Our Actions, 53

What Happened Next (Colombia, South America), 230

What Is a Rain Garden? (Slovakia), 146

What the Water Changed (Bundelkhand, India), 199

What the Water Teaches (Bundelkhand, India), 200

Who Is Fluent in the Language of Water (USA/ UK), 211

Why a Beaver? (USA/ UK), 207

Why Rain Gardens? (Slovakia), 147

Withdrawing from the Bank (Reaping Returns), 244

Women (Bundelkhand, India), The, 188

Yuzui or Fish Mouth Levee (鱼嘴) (China), The, 36

Zai Pits and Zero Runoff (Burkino Faso, Africa), 141

Zephaniah Phiri (Zimbabwe, Africa), 88

ABOUT THE AUTHORS

Philip Franses uses his expertise in holistic thinking and teaching to address global challenges through multistakeholder processes. He studied mathematics at New College, Oxford, and has designed intelligent software for a variety of organizations. Philip is the cofounder of The Flow Partnership, through which he creates platforms for communities to share their knowledge with each other, helping them restore water to its vital place within the cycles of nature. Philip also teaches holistic science and is the author of *Time, Light and the Dice of Creation: Through Paradox in Physics to a New Order*.

For more than 30 years, **Minni Jain** has been working with communities to regenerate their lives and landscapes. As cofounder and operations director of The Flow Partnership, she works to spread community-led, simple, successful, low-cost, traditional wisdom and methods of holding water and managing floods and droughts. To share and make available these community methods of landscape water resilience at a ground level, she has helped set up practical water schools in Africa, India, and Europe that operate both as online forums and as on ground community water hubs. She has helped cofound the Food Forest Fund to resource community projects globally. Minni was born and brought up in the Himalayas in India and now lives in the UK.